Technological Innovation

Strategy and Management

Technological Innovation

Strategy and Management

Juan Vicente García Manjón

Universidad Europea Miguel de Cervantes (UEMC), Spain

NEW JERSEY • LONDON • SINGAPORE • BEIJING • SHANGHAI • HONG KONG • TAIPEI • CHENNAI • TOKYO

Published by

World Scientific Publishing Co. Pte. Ltd.
5 Toh Tuck Link, Singapore 596224
USA office: 27 Warren Street, Suite 401-402, Hackensack, NJ 07601
UK office: 57 Shelton Street, Covent Garden, London WC2H 9HE

British Library Cataloguing-in-Publication Data
A catalogue record for this book is available from the British Library.

TECHNOLOGICAL INNOVATION
Strategy and Management

ISBN 978-981-121-145-4 (hardcover)
ISBN 978-981-121-146-1 (ebook for institutions)
ISBN 978-981-121-147-8 (ebook for individuals)

For any available supplementary material, please visit
https://www.worldscientific.com/worldscibooks/10.1142/11584#t=suppl

Desk Editor: Sandhya Venkatesh

Typeset by Stallion Press
Email: enquiries@stallionpress.com

ABOUT THE AUTHOR

Juan Vicente García-Manjón is a Professor of Innovation and Strategy in the undergraduate and MBA programs at the European University Miguel de Cervantes. Professor García-Manjón earned a B.S. in Business and a Ph.D. in Information Technology and Telecommunications from the University of Valladolid.

Professor García-Manjón developed his research in the field of innovation and technology. His doctoral thesis focused on the study of innovation models in the ICT sector. Likewise, his research analyzed the effects of R&D on business performance, the concentration of knowledge-intensive sectors (KIBS), the determination of innovation indicators for the study of innovation systems, and the design of public policies. His current research interests include strategic management of innovation and technology and leadership. He published his research in leading academic journals such as *Research Policy* and *Innovation: Management, Policy and Practice*. He also serves as an *ad hoc* reviewer for several journals in the field of innovation and technology, a program committee member for international conferences, and an R&D evaluator for industry agencies. He has experience as a technology-based start-up entrepreneur and business consultant and in managing non-profit organizations and technological centers.

ACKNOWLEDGMENTS

The idea to write this book arose from a willingness to share my experience in research and professional practice in the field of innovation and technology. I am thankful to World Scientific Publishing Co. Pte. Ltd. (WSPC) for giving me the opportunity to turn it into a reality. I want to thank my editors, especially Sandhya Venkatesh, for their support and patience in the difficult process of writing this book.

Likewise, I am grateful to Dr. Brett Trusko, President and CEO of the International Association of Innovation Professionals (IAOIP), for his willingness to write the foreword for this book and for his guidance and advice.

I am especially thankful to Nina Fazio, who helped me with the proofreading and initial editing of this book. She also became one of my main supporters in this long and hard writing process.

I also want to thank the management of my university, Universidad Europea Miguel de Cervantes, who encouraged me to approach the project and provided me with the flexibility necessary to complete it, with special mention to all my colleagues who provided helpful suggestions on the content and book's structure.

Finally, my most meaningful thanks and gratitude go to Isabel, Álvaro, and Maria; and to my parents, Juan and Elisa. They all are the true assets of my life.

FOREWORD

By Professor Brett Trusko

In Chapter 2 of his important new book, author Juan Vicente García Manjón quotes Usher (2011): "The process of innovation has frequently been held to be an unusual and mysterious phenomenon of our mental life. It has long been regarded as a result of a special man of genius." Professor García Manjón goes on to state that this imagined "magical character" of the innovation stream makes it difficult to view innovation as a process that can be divided into different stages and consequently be organized and managed.

There are tens of thousands of consultants selling the "magical character" of innovation, based upon some process they, and they alone, understand — which is, of course, an oxymoronic statement because a process is defined as "a series of actions or steps taken in order to achieve a particular end." Or, in other words, something repeatable and predictable. However, organizations that understand and follow prescribed processes as outlined in the Oslo Manual and the developing standards of ISO 279 (56000 series) find that their innovation horizons are less variable and volatile than those who do not.

Professor García Manjón has written a wonderfully researched and readable book that will largely demystify innovation. This book can be viewed either as a first-class documentation of the contemporary literature on innovation or as a stand-alone book that would be easy to read by non-academics.

For example, early in the book, Professor García Manjón does a thorough job of outlining key definitions and concepts related to innovation, utilizing the Oslo Manual as a basis on which to build, in a way that anyone who reads this book will be educated enough on the developing world of innovation standards that they can be fluent in the most important concepts in innovation and their interrelationships with each other. Given that governments and large companies worldwide are seriously considering the adoption of the ISO 56000 standards as they continue to develop, this book will likely prove to be an important book in understanding the standards as well as the history and basis of those standards.

From the perspective of technology innovation, one must understand that the entire universe goes through lifecycles. With regard to technology, we happen to be in a stage we could refer to as adolescence. Much like a 16-year-old is likely to be a more reckless driver than a 50-year-old, the use or misuse of information technology via social media and hacked credit cards are just a sign of a maturing industry. And just like a 16-year-old driver, it is exciting and full of life. For those who are unnerved by the changes, rest assured that books such as this will help us all reach a better place by getting control of our innovations and the resultant technology.

Professor García Manjón discusses the strategy of innovation as well as its sources, both from inside and outside the organization. He discusses the tools, patterns, and sources of technological innovation and how they all fit together as a functional whole.

He spends time with and explains the fundamentals of leadership and organization as well as developing a culture. Because he has so meticulously researched the topic, his findings and suggestions are tied back to the best research done in this space — research that has been validated and shown to be effective time and time again.

He has dedicated a chapter to the importance of human capital, creativity, and learning. Anyone familiar with innovation will quickly discover that the linchpin to innovation is a robust construct of people and culture. Drucker is often quoted as saying, "Culture eats strategy for lunch." As an experienced innovation professional, Drucker was prescient to note this 30 or 40 years ago, and it is even more important that Professor García Manjón has dedicated a part of his book to this premise.

Finally, Professor García Manjón has also done an outstanding job with several chapters that discuss the process of innovation, the formulation of teams, and the monitoring and ongoing management of the innovation process. His attention to the professionalization of innovation through process, fully supported by some of the greatest writers on innovation and business as well as the remarkable work of the Oslo Manual, means that any serious innovation professional will need this book on their bookshelf — or, more accurately, in their e-reader.

Innovation is not easy, and it is sometimes ugly. As I like to say, "The most dangerous job is the Chief Innovation Officer." While I don't have statistics, anecdotal evidence I have acquired from working with major companies and countries from around the world suggest that innovation is not well understood by most CEOs, and because of the difficulty in producing new innovations on a schedule, it is likely that a Chief Innovation Officer's job is at risk anytime the economy is in a downturn. This isn't necessarily a problem for most innovation professionals as they thrive in times of change, but what if you like your job? The best way to mitigate your risk is to read this book carefully and become a leader in the professionalization of this exciting discipline.

Professor Brett Trusko
President and CEO
The International Association of Innovation Professionals
Secretariat, US TAG, ISO 279/56000
Assistant Professor, Texas A&M University (USA)
Distinguished Professor
Symbiosis Institute of Business Management (India)

CONTENTS

CHAPTER 1

UNDERSTANDING INNOVATION AND TECHNOLOGY: KEY DEFINITIONS AND CONCEPTS

1.1 THE NEW ECONOMIC PARADIGM AND THE IMPORTANCE OF TECHNOLOGY AND INNOVATION

Along the history of humankind, technology has supplied a cornerstone for social and economic progress. Technology has triggered productivity, improved the quality of life, paved way for new organizational forms, and promoted human interaction processes. Technology has also provided numerous breakthroughs that have led societies to amend the way in which they see the world and interact with it. Societies have evolved throughout different paradigms that have eased the understanding and interpretation of their environment.

To shed light onto the comprehension of the role of technology and innovation nowadays, we have to delve into the understanding of the term "paradigm." The term was coined by Kuhn in his work *The Structure of Scientific Revolutions*, where he studied the development of science along the course of history [Kuhn, 1962], and it has been widely used by scholars and researchers in a variety of scientific fields, including economic theory.

More precisely, Dosi, based on Kuhn's work, referred to the expressions "scientific paradigm" and "technological paradigm." A "scientific paradigm" is described by the author as "an outlook which defines the relevant problems, a model and a pattern of inquiry," while a "technological paradigm" is defined as a "model and a pattern of solution of selected technological problems," which denotes an agreement among the participants on what is to be considered an improvement of a product, service, or technology. Additionally, on the ground of a given "technological paradigm" and as the pattern of normal problem-solving, technologies can progress and evolve, following what Dosi coined as "technological trajectories" [Dosi, 1982, p. 152]

Technological paradigms have been broadly used in the study of the economics of technological change in order to explain how technology breakthroughs are the basis for production systems and value creation. Thus, according to Pérez [2003, p. 5], the term paradigm has broadened, considering it "a collectively shared logic at the convergence of technological potential, relative costs, market acceptance, functional coherence, and other factors." This leads to the concept of a "techno-economic paradigm," which highlights the symbiosis between economics and technology. The techno-economic approach represents a group of technologies and solutions that are constantly being improved with a great effect on the economic system itself [Pérez, 2003].

Consequently, technology and economy seemingly interact with each other in order to compose an assembly of technical procedures, new sets of knowledge, and new practices of value creation that constitute the basis for an explicit economic activity model.

1.1.1 *How do Science and Technology Evolve Over Time?*

We are conscious that economic models and technological assets evolve and change over time, leading to new technological solutions and economic applications. In this vein, we are interested in exploring how technological and economic systems change and finding the tipping point for a technological and economic shift.

To clear up the question, we refer to the works of those authors who tried to explain economic evolution for many years. The first mention is to Kondratieff [1892–1938] who, throughout his work, *The Long Waves in Economic Lives*, argues for the existence of economic cycles of about 50 years in length. According to the author, these cycles are characterized by the formation of capital goods at the beginning of the cycle, followed by an increase in consumption and employment. Afterward, the decline of investment drives the economy to high unemployment rates and overcapacity in the use of capital goods, which after a period of time leads to new capital formation and the commencement of a new cycle. However, Kondratieff calls into question the role of technology as explaining or causing the shifts in economics, and prioritizing the role of the market to trigger the use of new production techniques. Thus, the author states that "scientific-technical inventions in themselves, however, are insufficient to bring about a real change in the technique of production. They can remain ineffective so long as economic conditions favorable to their application are absent" and complements "changes in technique have without doubt a very potent influence on the course of capitalistic development. But nobody has proved them to have an accidental and external origin" [Kondratieff, 1935, p. 112]. Therefore, although the author recognizes the role of technology, he does not consider it a central element in economic change.

We have to go to the fourth decade of the last century to come up with Schumpeterian theories, which hold that disruptive technological innovations lead to economic development, recognizing the key role of technology in the economic shift. Schumpeter coined the term "Business Cycles" [Schumpeter, 1939, p. 150], which can be traced back to Kondratieff's long waves. This vision of technology as a trigger for economic growth can be observed in the following citation from the author: "If innovations are being embodied in new plant and equipment, additional consumers' spending will result practically as quickly as additional producers' spending. Both together will spread from the points in the system on which they first impinge, and create that complexion of business situations which we call prosperity."

So, a techno-economic paradigm provides the society with a set of assumptions, rules, techniques, solutions, ways of producing, etc., that outline the main research challenges and a collection of acceptable and proved findings that represent the state of the art for that specific paradigm. The question, as we stated before, is how long will this paradigm be acceptable? Well, it all depends on the accumulated shortcomings and its sustainability. If the existing paradigm is no longer sustainable, it will be replaced by a new one.

In this vein, and in order to comprehend the way the shift occurs, it is worth coming back to Kuhn's vision. The author argued that scientific shifts progress through the following stages [Kuhn, 1962] (Table 1.1):

Table 1.1. Different Stages in Scientific Shift.

Phases	Explanation
Normal science	For the author, normal science means "research firmly based upon one or more past scientific achievements, achievements that some particular scientific community acknowledges for a time as supplying the foundation for its further practice." Normal science constitutes a scientific paradigm when its achievements fulfil two criteria; being sufficiently extraordinary to entice an enduring group of adherents away from competing modes of scientific activity and at the same time, it is open-ended to leave all sorts of problems for the redefined group of practitioners to resolve [Kuhn, 1962, p. 10].
Normal science as "puzzle-solving"	This concept relates to the fact that, in many cases, the adherents to a paradigm choose mostly problems that are expected to have solutions. To a great extent, the author adds, "these are the only problems that the community will admit as scientific or encourage its members to undertake" [Kuhn, 1962, p. 37].
Anomaly and the emergence of scientific discoveries	This relates to the dysfunction of normal science when shortcomings and pitfalls appear failing to explain certain phenomena under the existing paradigm. However, the author poses that real discoveries only commence with the recognition of anomalies, which in some way contravene the existing paradigm of normal science and closes only when the paradigm theory has been adjusted so that the anomalous has become the expected [Kuhn, 1962, p. 52].
Crisis and the emergence of scientific theories	The aforementioned discoveries lead to a change in the scientific paradigm. In order to place new discoveries in the existing paradigm, it is necessary to discard some previous standard beliefs and replace them with new ones. However, when the existing scientific paradigm is unable to explain adequately the newly studied phenomena, the paradigm goes into crises.
The Nature and necessity of scientific revolutions	When the old paradigm cannot be revitalized, the community searches for a new paradigm to replace the former one. With a new paradigm in mind, scientists adopt new instruments and look in new places. The paradigm's shift leads the scientific community to see the world differently, under the prism of the new principles set by the new paradigm.

Source: Kuhn [1962].

1.1.2 *Major Technological Revolutions Along the Course of History*

Once we have analyzed how science and technology evolve, we would like to examine which have been the major technological shifts along the course of history and how they happened. Carlota Pérez, a Venezuelan scholar who studies the evolution of technology and economic development, refers to the term "technology system" as a bunch of interconnected, individual innovations, which in turn are interconnected in technological revolutions. The author explains that the appearance of new technologies is not a random or isolated event, since it involves other agents of change, such as providers, distributors, or consumers according to the concept of Schumpeterian clusters, which are the results of techno-economic and social interactions between producers and users within complex, dynamic networks [Pérez, 2003]. Similarly, Pérez highlights the importance of incremental innovations in the evolution of technological solutions following breakthrough or radical innovations because although radical innovations play a key role in determining new investments and economic growth, technological evolution depends on incremental innovations [Pérez, 2003].

Coming back to the term "technological revolution," it can be defined as "a set of interrelated, radical breakthroughs, forming a major constellation of interdependent technologies; a cluster of clusters or a system of systems" [Pérez, 2004]. The author adds that a technological revolution is distinguished from a simple random collection of technology systems by two basic features: a strong interconnectedness and interdependence of the participating systems in their technologies and markets, and the capacity to profoundly transform the rest of the economy and eventually the society. The author posits the existence of five major technological revolutions that are summarized in Table 1.2.

The first technological revolution, known as the "Industrial Revolution," took place in 1771 in Britain. The starting point was the opening of Arkwright's mill in Cromford, the first and most important cotton mill established by Sir Richard Arkwright, who pioneered the development of his frame-spinning machine, revolutionizing the manufacturing of fabrics, and constituting a cornerstone for the Industrial Revolution. According to Pérez [2003], this technological revolution set up new technologies, such

Table 1.2. The Five Successive Technological Revolutions, 1770–2000.

Technological revolution	Popular name for the period	Core country or countries	Big-Bang initiating the revolution	Year
First	The "Industrial Revolution"	Britain	Arkwright's mill opens in Cromford.	1771
Second	Age of steam and railways	Britain (spreading to the continent and USA)	The "Rocket" steam engine for the Liverpool–Manchester railway is tested.	1829
Third	Age of steel, electricity, and heavy engineering	USA and Germany forging ahead and overtaking Britain	The Carnegie Bessemer steel plant opens in Pittsburgh, Pennsylvania.	1875
Fourth	Age of oil, the automobile, and mass production	USA (with Germany at first vying for world leadership), later spreading to Europe	First Model-T comes out of the Ford plant in Detroit, Michigan.	1908
Fifth	Age of information and telecommunications	USA (spreading to Europe and Asia)	The Intel microprocessor is announced in Santa Clara, California.	1971

Source: Pérez [2003].

as mechanization of the cotton industry, fabrication of wrought iron, and the intensive use of machinery. At the same time, it created or redefined new infrastructures such as canals and waterways, turnpike roads, and water power throughout new and improved water wheels. The paradigm of new techno-economic innovation principles was based on the factory as the center of production, mechanization, the importance of productivity, and the existence of local networks.

The second major shift into a new economic paradigm started in 1829 and was known as the "Age of steam and railways." According to Pérez [2003], the new technologies involved in this technological revolution were the steam engines and new machinery, iron and coal mining, railway construction, rolling stock production, and steam power. Accordingly, there appeared new infrastructures such as the railway, telegraph, universal postal service, functional ports, depots, worldwide sailing ships, and city gas. The new techno-economic paradigm changed, provoking the agglomeration of economies around industrial cities and the conformation of national markets, the appearance of power centers with national networks, the importance of scale, standardization of production parts against the handicraft economy, and the growing interdependence of machines and means of transportation.

The author referred to the third technological revolution as the "age of steel, electricity, and heavy engineering." The starting point was the opening of the Carnegie Bessemer steel plant in Pittsburgh, Pennsylvania (USA) in 1875, moving the gravity center of technological expansion to the USA and Germany and later to Britain. The new technologies that appeared with this technological shift were cheap steel, steam engines for steel ships, heavy chemical and civil engineering, the electrical equipment industry, copper and cables, canned and bottled food, and paper and packaging. These new technologies allowed major shifts in infrastructures such as worldwide shipping, transcontinental railways, great bridges and tunnels, worldwide telegraphy, mainly national telephone networks and electrical networks.

Continuing reference to the work of Pérez, the fourth technological revolution is known as the "age of oil, automobiles, and mass production." The starting point for this age is when the first "Model-T" came out of the Ford plant in Detroit, Michigan (USA) in 1908. The technologies associated with this technological area include mass production throughout the assembly line, cheap oil-based fuels, petrochemicals, internal combustion engines, electrical home appliances, and refrigerated and frozen foods. All these technologies were accompanied by changes in infrastructures, such as networks of roads and highways, oil pipelines, universal electricity, and analog telecommunications (telephone, telex, and cablegram).

Finally, Pérez refers to the fifth technological revolution as the "age of information and telecommunications," which originated with the announcement of the first Intel microprocessor in California (USA) in 1971. This era is characterized by what is called the "information revolution," and it is based on microelectronics, computers, software, telecommunications, control instrumentation and computer-aided biotechnology, and new materials. These technologies triggered the development of infrastructures for digital telecommunications, the Internet, electrical networks, and high-speed, multi-modal physical transport links.

1.1.3 *The Dominant Techno-Economic Paradigm Nowadays*

What is the predominant techno-economic paradigm today? Are we still under the influence of the fifth technological revolution, that is, the age of information and telecommunications?

As we mentioned earlier, the irruption of information and communication technologies (ICTs) can be traced back to 1971 when Intel introduced the first commercially viable microprocessor, which made it possible to incorporate all of the functions of a central processing unit (CPU) onto a single integrated circuit [Knell, 2010]. This kind of technology made possible the creation of the first personal computers, software, and integrated circuits in a variety of new products and services. And later, it facilitated the development of digital telecommunication networks and the Internet, which constituted one of the biggest shifts in history ever. Likewise, ICTs have experienced exponential growth in their information processing capacity, which leads to still greater computing capacity today at the same levels of investment as years ago [Biagi, 2013].

However, according to Hanna [2010], we are still in the early phase of a long-term technological wave and productivity revolution with the use of information and telecommunication technologies. These technologies are expected to produce intense decline in prices and increases in ICT system performance and intelligence. Hanna holds that the ICT revolution represents a techno-economic paradigm shift with deep implications for the regeneration of productive and institutional structures in developed and developing countries alike. He also maintains that "the ongoing technological revolution is so profound and pervasive that it challenges many traditional economic concepts that are rooted in incremental thinking" [p. 29].

At the same time, it is widely assumed that the implementation of ICTs leads to improvements in productivity at both the national and business levels. This assumption has been thoroughly studied in the literature, even though at the beginning there were some voices that claimed that ICT and productivity were not clearly connected. Thus, Robert Solow stated, "You can see the computer age everywhere but in the productivity statistics" [Solow, 1987]. Fortunately, this statement was not exactly true, and it is a known fact that as long as ICTs develop, so will efficiency and productivity. In fact, there have been a large number of authors studying and concluding positive correlations between ICT and productivity. Accordingly, Biagi [2013] in the report "ICT and Productivity: A Review of the Literature" argues that ICTs positively influence growth and productivity. This influence can be direct, since growing productivity in the ICT industry has a direct and proportional consequence in terms of its weight on the GDP, on aggregate productivity [Jorgenson *et al.*, 2008; Gordon, 2000, 2012; van Ark *et al.*, 2008]. Conversely, this relationship can have indirect effects as well, since ICT plays an important role in other industrial or service sectors. Particularly, the same report points out that ICTs are enablers of product, process, and organizational innovations in other sectors that implement the use of ICT. Similarly, these technologies facilitate the management, storage, and transmission of information, helping to reduce market failures due to information asymmetries.

So, ICT seemingly impinges on productivity and growth. Nevertheless, ICT's impact is so pervasive that some authors qualify them as a General Purpose Technology (GPT) [Jovanovic and Rousseau, 2005]. A GPT is a "technology that initially has much scope for improvement and eventually comes to be widely used, to have many uses and to have many Hicksian and technological complementarities" [Lipsey *et al.*, 1998, p. 43]. A technology must fulfill some criteria to be considered a GPT, such as pervasiveness (the GPT should spread to most sectors); improvement (simultaneous increase in performance and cost reduction) and capacity to spawn innovation (easing the invention and production of new products and processes) [Trajtenberg and Bresnahan, 1992]. There are authors like Jovanovic and Rousseau [2005, p. 1186] who compare electricity and ICTs in terms of their ability to generate economic growth. They conclude that "the evidence shows similarities and differences between the electrification and the IT eras. Electrification was more pervasive, whereas IT has a clear lead in terms of improvement and innovation spawning." This ability to induce improvements in other economic sectors, mainly the productive one, and the decline in ICT prices make the authors hopeful about the growth forecasts for ICTs.

Nobody doubts that ICT has a great impact on many different fields. For instance, ICT has a clear effect on governance through the decentralization of power, easing information flow (online governmental information, freedom of information requests), fostering new types of communities, and provisioning of different roles for the government (citizen involvement, policy formulation, and legislative branch) [Hanna, 2010].

Likewise, ICT widely impinges on education. Hanna [2010] argues that with the rapid pace of technology shifts, it is an imperative for a well-educated and skilled workforce, new ways of knowledge sharing, and lifelong learning. Consequently, ICT is crucial to limit the fast-increasing costs of education, and he adds that technology-enhanced learning will require substantial innovation in the education sector. So, ICT has the potential to transform how people learn throughout their lives [Resnick, 2002], while distance learning is expanding the learning ecosystem beyond schools and enabling new types of "knowledge building communities" [Selinger, 2004]. Hanna, citing the work of Goldin and Katz [1998] and de Ferranti *et al.* [2002], maintains that technology and skills play critical and balancing roles in increasing productivity and in ultimately increasing the economic growth. Through the use of ICT, the education process has become more inclusive, accessible, and cost-effective. In this vein, it is worth citing the experience of the so-called Massive On-line Open Courses (MOOCs), which make leading teachers accessible to a huge and dispersed number of students worldwide. Nobody doubts that ICT is transforming the educational arena, including its business models, making education more accessible and cost-effective, and reinventing the way we learn and develop new skills.

Similarly, the health sector is also changing quickly due to the implementation of ICT, even though future potential is still huge. ICT, software, new telecommunication networks, and mobile communications foster health education and training and enable new diagnostic systems, telemedicine and telecare, patient information, and medical records management [Hanna, 2010]. According to the report "Improving Health Sector Efficiency: The Role of Information and Communication Technologies" [OECD, 2010a], ICT implementation has four broad and interrelated effects on healthcare systems, such as an increase of quality of care and efficiency, a decrease in clinical services operating costs, a decrease in administrative costs, and the capacity to enable totally new modes of care. However, one of the most pervasive effects of ICT is on the business sector. According to Hanna [2010], strategic information systems (electronic trading and financial payment clearance and settlement systems) are key to economic competitiveness and operations in the global economy. Such systems, adds the author, represent the new national infrastructure of the knowledge economy, which require substantial investment and reinforce other economic activities that bring important spillover effects to competitiveness in the private sector. Concerning the business sector, the effects of ICT on business are pervasive, thereby changing the ways of managing, marketing, and connecting with customers and networking with other businesses. The establishment of ICT-based business management systems (CRMs, ERPs, and SCMs among others) is increasing firm competitiveness, enabling new ways of relating with suppliers, customers, and the competition. Especially relevant are the new e-commerce systems and platforms, which have dramatically changed consumer habits and behaviors and enabled new marketing channels to reach new clients, where geographic distances become irrelevant. Similarly, mobile communications are having pervasive effects on all sectors and economic activities, disrupting traditional business models, creating new ways of value formation, and facilitating new economic activities. Particularly relevant have been the changes in the so-called sharing-economy. The sharing-economy is an economic and social activity grounded on the idea that underutilized assets, services, or capacities can be shared directly by individuals on a free or for-fee basis [Botsman,

2015]. This kind of economic activity is exemplified in companies like Uber or Airbnb, which have transformed the business models of traditional industries such as transport and hospitality. However, the key enabling element of this new economic model is ICT-based platforms that allow individuals and consumers to get connected and support the service delivery.

Beyond the consideration of ICT as a vehicular technology to drive economic growth and to boost productivity in many sectors, we want to point out that the economic and productive systems in the most developed countries are characterized by an intense use of knowledge as the main input of the creation of value. ICT is transforming the ways in which government, business, and society manage the fundamental productive factor, which is knowledge. Consequently, the term "knowledge-based economy" has been coined to describe an economic system with a great dependence on knowledge, the massive use of information, and the demand for a high-skilled workforce. A knowledge-based economy is characterized by the intensive use of knowledge or new technologies, also it is an economy where all sectors are knowledge intensive, where there is a high responsiveness to new ideas, technological change, innovativeness, and the use of highly skilled workforce and continuous learning [Smith, 2000]. There is evidence that innovation is a leading factor in economic growth at both the national and international levels and that ICT is quickening and disseminating innovation. Also, at the firm level, research and development favor the capacity to absorb and make use of knowledge of all types [OECD and Eurostat, 2005] Correspondingly, technological change has been acknowledged as the single most important contributing factor to long-term productivity and growth [Grübler, 2003]. The concept of the "knowledge economy" entails different approaches including the rise of new science-based industries, increased knowledge intensity in certain industries, and the role of continuous learning and innovation in virtually all businesses [Powell and Snellman, 2004].

Hanna [2010, p. 51] qualifies ICT as a "powerful enabler of innovation" and maintains that "its application to research, design, services, logistics, finance, marketing, and learning has enabled enterprises to become more efficient, flexible, and innovative, through process innovation, product and service innovation, and the creation of new business models. Information and communication activities are at the heart of the innovation process, and ICT has become a tool for amplifying brainpower and for innovation."

As stated earlier, the qualification of ICTs as GPTs according to Bresnahan and Trajtenberg [1995] leads us to consider ICT as pervasive, a technology that allows continuous improvements and experimentation and facilitates innovation. In the same vein, we can report some indirect effects of ICT investment, discovering that these technologies are enablers of product, process, and organizational innovations, many times through co-invention [Biagi, 2013].

Nowadays, the economic arena is characterized by the continued economic growth demands from policy-makers and industries. The increase in global competition, the existence of shorter product life cycles, the appearance of pervasive new technologies, customization on the one hand and high commoditization of products and services on the other hand, and price sensitiveness all depict the landscape of markets today. Accordingly, organizations are asked to pinpoint and assimilate new knowledge and capabilities that are often available outside their own borders, since today a comprehensive range of knowledge from different bases is required to develop new products, services, or processes [OECD, 2010b]. The necessity to increase the pace of innovation is a major motivation to involve external sources such as research institutes, companies, and related markets [OECD, 2008]. Therefore, apart from the fact that more players are involved in the innovation process, we also have to consider that innovation is occurring through interactive and collaborative processes [OECD, 2010b]. This new approach to innovation would not be possible without the Internet and ICTs in general. According to Hanna [2010], the Internet favors the creation of networks of knowledge throughout inventors, scientists, and innovative firms, through which it is possible to harness the capabilities and resources needed to drive the innovation process inside the organization. All of the above stem from the concept of "open innovation," a term that was coined by Chesbrough [2003] to describe a new paradigm of innovation in which collaboration, a multidisciplinary approach, openness, and global scope conform to the new way of innovating.

> *At its root, open innovation is based on a landscape of abundant knowledge, which must be used readily if it is to provide value for the company that created it. However, an organization should not restrict the knowledge that it uncovers in its research to its internal market pathways, nor should those internal pathways necessarily be constrained to bringing only the company's internal knowledge to market.* [Chesbrough, 2003, p. 37]

1.1.4 *Which is the Next Technological Revolution?*

Up until now, we have examined the role of science and technology in societal and economic evolution. Equally, we have referred to the major Industrial Revolutions so far and how they have brought profound economic and societal changes. Apparently, the Internet and the pervasiveness of ICT are transforming business and societies. Consequently, it is reasonable to envision future technological shifts and how they will impact our socioeconomic ecosystems. In 2003, the National Science Foundation (NSF) published the report "Converging Technologies for Improving Human Performance" [Roco and Bainbridge, 2003] in which they set the basis for the analysis of evolution in science and technology (Fig. 1.1).

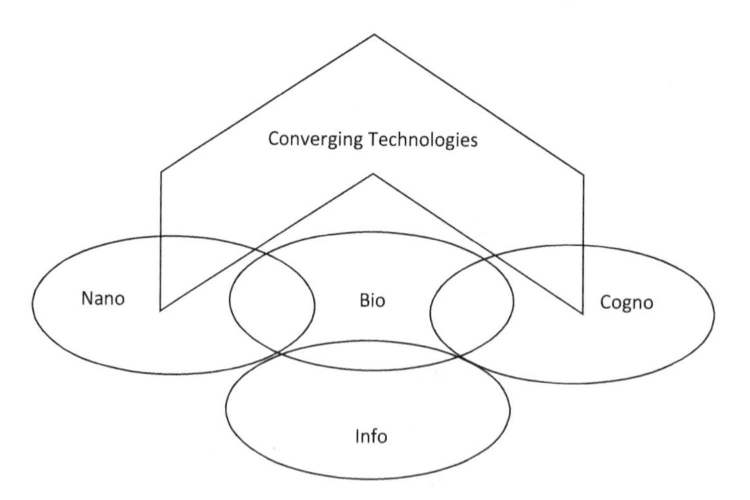

Figure 1.1. Converging Technologies.

Source: Own elaboration based on Roco and Bainbridge [2003].

The report foresees a major shift in science and technology in the first decade of the 21st century in the form of a synergetic combination of four scientific areas, encompassing nanotechnology, biotechnology, information technology, and new technologies based in cognitive science — known with the acronym NBIC. According to the report, this shift will dramatically impact human abilities, societal outcomes, productivity, and the quality of life.

As mentioned in the report, this scientific and technological convergence will "improve work efficiency and learning, enhancing individual sensory and cognitive capabilities, revolutionary changes in healthcare, improving both individual and group creativity, highly effective communication techniques including brain-to-brain interaction, perfecting human machine interfaces including neuromorphic engineering, sustainable and 'intelligent' environments including neuro-ergonomics, enhancing human capabilities for defense purposes, reaching sustainable development using NBIC tools, and ameliorating the physical and cognitive decline that is common to the aging mind" [Roco and Bainbridge, 2003, p. 9].

Correspondingly, Pérez [2003, p. 13] highlights that in the early 2000s, biotechnology, bioelectronics, and nanotechnology seemed to be at "a stage equivalent to the oil industry and the automobile at the end of the nineteenth century or to electronics in the 1940s and 1950s, with vacuum-tube TVs, radar and analog control equipment and telecommunications" and stressed the unpredictable character of the power hidden in the forces of life and the infinitely small. Some authors contend that nanotechnology could qualify as a new GPT [Knell, 2010], due to the wide variety of applications in all kinds of sectors.

Also, the futurist and inventor, Ray Kurzweil, in his book *The Singularity is Near* [Kurzweil, 2006], predicted a radical evolution of technologies in the next decades. As reported by the author, we will live the most profound and pervasive changes in the history of humanity. Kurzweil argues that there will be a genetic, nanotechnology, and robotic revolution. At the same time, the author theorizes that these advances will impact the field of the human brain and consequently drive the development of artificial intelligence (AI). In an article in *Forbes* magazine [Satell, 2016], Kurzweil maintains that there are three reasons to think that the "singularity" is near. The first one is that we have gone beyond Moore's law, the famous prediction of Intel co-founder Gordon Moore in which he suggests that the number of transistors on a microchip would double about every 18 months. Second, robots are doing human jobs, including those in the creative realm. And third, we are editing genes to control diseases.

The irruption of the next technological revolution will depend on "when the current information revolution approaches the limit of its wealth-generating power" [Knell, 2010, p. 137].

To understand how a new technological revolution is born, we have to refer again to the work of Carlota Pérez, who distinguishes two periods in each technological revolution — an installation and a deployment phase — with a turning point in the middle characterized as a bubble breakdown, followed by a recession. The author argues that the flow of development that stems from each technological revolution lasts half a century or more [Pérez, 2016]. Accordingly, Pérez argues that in this specific moment of the current shift, we have already seen the initial impact of ICT, which has transformed entire industries and opened the way to new opportunities — from turning tangible products into services to the creation of the home office and the globalization of production and trade. The author maintains that ICT has also changed some of the ways of consumption, favoring a greater use of information, being more innovative and entrepreneurial, and based on the use of networks and platforms. Yet, as reported by the author, this transformative work is far from done, since the ICT revolution "has the capacity to facilitate wide-ranging sustainable innovations to radically reduce materials and energy consumption while stimulating the economy. It can significantly increase the proportion of services and intangibles in GDP as well as in lifestyles" [Pérez, 2016, p. 192].

Similarly, in 2009, the European Commission studied the situation of the so-called "key enabling technologies" (KETs), which was presented in a communication titled "Preparing for Our Future: Developing a Common Strategy for KETs in the EU" and assumed that these technologies would be crucial for the future competitiveness of the EU. This document analyzed the situation of KETs in Europe and pointed to advanced materials, nanotechnology, micro- and nanoelectronics, industrial biotechnology, and photonics as the major technologies that will lead to an increase in productivity and employment in the future.

One year later, in 2010, the European Commission sponsored the report "European Competitiveness in Key Enabling Technologies: Summary Report" [Aschhoff *et al.*, 2010, p. 28]. KETs are defined as "new technologies that enable product and process innovation in manufacturing." In general, applying KETs will enable producers to use labor, capital, energy, and other inputs more efficiently. The report argues that there have been certain technologies that throughout history have dramatically impelled innovation and technical progress, leading to higher levels of productivity and allowing the appearance of breakthrough innovations, naming these technologies as "key enabling technologies." Specifically, according to the report, KETs can support increases in productivity, more efficient use of production factors, and technical progress in the production function. Likewise, KETs in R&D will produce a stock of knowledge that could affect productivity in other economic sectors and accelerate technical progress.

1.2 UNDERSTANDING SCIENCE, RESEARCH, AND TECHNOLOGICAL DEVELOPMENT

The distinction between science and technology has been a topic widely discussed in the academic and business world. According to Brooks [1994, p. 477], "the debate about science and technology policy has been implicitly dominated by a 'pipeline' model of the innovation process in which new technological ideas emerge as a result of new discoveries in science and move through a progression from applied research, design, manufacturing and, finally, commercialization and marketing." Likewise, the author argues that one concern about both terms is the confusion that exists between "science" and "engineering," which stems from an undue worry about the originality and priority of the conception phase in order to guarantee successful technological innovations. The foregoing leads to equating the research and development process with the innovation process itself. However, the author calls this into question, affirming that even though science, technology, and innovation are highly reliant, they are different from each other.

To shed light on the relationship between science and technology, it is worth citing the report "Science: The Endless Frontier," which was commissioned by President Roosevelt. The report calls for a centralized approach to science sponsored by government and defends the key role that basic research plays in leading to new knowledge, highlighting that this creates the basis for practical applications of it [Bush, 1945, p. 13]. In the same vein, Bunge [1966, p. 329] defines technology as applied science and posits "the method and the theories of science can be applied either to increasing our knowledge of the external and the internal reality or to enhancing our welfare and power. If the goal is purely cognitive, pure science is obtained; if primarily practical, applied science." Conversely, we can consider that technology also feeds new scientific possibilities, since technological infrastructure and instrumentality (advances in instrumentation and experimental techniques) are necessary to promote scientific development, and advances in these technologies lead to new science [Price *et al.*, 2013; Pitt, 1995].

Other authors argue that science and technology are independent realms and look for a new interpretation of technology. Thus, Layton states that, even assuming that science and technology are quite close communities, "each community has its own social controls — such as a reward system — which tend to focus the work of each on its own needs. These needs determine not only the objects of concern, but the language in which they are discussed" [Layton, 1971, p. 565].

However, after the Industrial Revolution and, according to constructivist models, science and technology were fused into a single composite, a single entity named "the science," "technology complex," or "technoscience." According to Giere [1993], science can be described as a complex system, composed of actions and agents, aiming at the production of models to depict reality. In this same vein, the author characterizes technology as a complex system of actions aimed at producing technological processes or artifacts.

Science evolves through the creation of new general and applied knowledge to the state of the art in one specific field. To do so, research and experimental development are key elements. In order to delve into the analysis of those concepts, it is worth citing the OECD's "Frascati Manual: Guidelines for Collecting and Reporting Data on Research and Experimental Development," which includes the definitions of basic concepts, data collection guidelines, and classifications for compiling R&D statistics. According to the F*rascati Manual* [OECD, 2015], R&D only represents one of the stages of the innovation process. Hence, research and experimental development comprise the creative work undertaken in a systematic way to increase the body of knowledge, including that of the man, culture, and society and the use of that knowledge to create new applications. Following the explanation of the R&D concept and according to the F*rascati Manual*, R&D is composed of three subcomponents: basic research, applied research, and experimental development. To be considered R&D, the activity must be novel, creative, uncertain, systematic, and transferable and/or reproducible. The F*rascati Manual* defines "basic research" as "experimental or theoretical work undertaken primarily to acquire new knowledge of the underlying foundations of phenomena and observable facts, without any particular application or use in view" [OECD, 2015, p. 50]. "Basic research analyses properties, structures and relationships with a view to formulating and testing hypotheses, theories, or laws. The reference to no particular application in view of the definition of basic research is crucial, as the performer may not know about potential applications when doing the research or responding to survey questionnaires" [OECD, 2015, p. 50]. Furthermore, according to the aforementioned manual, "the results of basic research are not generally sold but are usually published in scientific journals or circulated to interested colleagues. Occasionally, the publication of basic research may be restricted for reasons of national security" [OECD, 2015, p. 50]. Furthermore, we can distinguish between two kinds of basic research as follows:

- Pure basic research, which is carried out for the advancement of knowledge, without seeking economic or social benefits or making an active effort to apply the results to practical problems or to transfer the results to sectors responsible for their application.

- Oriented basic research, which is carried out with the expectation that it will produce a broad base of knowledge likely to form the basis of the solution to recognized or expected current or future problems or possibilities.

Going one step further, the F*rascati Manual* also defines applied research "as the original investigation undertaken in order to acquire new knowledge. It is, however, directed primarily toward a specific, practical aim or objective." The F*rascati Manual* takes the following considerations about applied search [OECD, 2015, p. 51]:

- Applied research is undertaken either to determine the possible uses for the findings of basic research or to determine new methods or ways of achieving specific and predetermined objectives.
- It involves considering the available knowledge and its extension in order to solve actual problems.
- The results of applied research are intended primarily to be valid for possible applications to products, operations, methods, or systems. Applied research gives operational form to ideas.
- The applications of the knowledge derived can be protected by intellectual property instruments, including secrecy.

Finally, the manual defines experimental development as "systematic work, drawing on knowledge gained from research and practical experience and producing additional knowledge, which is directed to producing new products or processes or to improving existing products or processes" [OECD, 2015, p. 51]. Experimental development is just one possible stage in the product development process: that stage when generic knowledge is actually tested for the specific applications needed to bring such a process to a successful end. During the experimental development stage, new knowledge is generated, and that stage comes to an end when the R&D criteria (novel, uncertain, creative, systematic, and transferable, and/or reproducible) no longer apply.

Summarizing the main concepts that have been studied, we have Table 1.3 in which we detail the objectives, content, and expected results from each of the stages in the R&D process.

Table 1.3. Concepts of Basic Research, Applied Research, and Experimental Development.

Typology	Objectives	Content	Results
Basic research (BR) [p. 50]	Acquisition of new knowledge of the underlying foundations of phenomena and observable facts, without any particular application or use in view.	BR analyses properties, structures, and relationships with a view to formulating and testing hypotheses, theories, or laws.	The results of basic research are not generally sold but are usually published in scientific journals or circulated to interested colleagues. The main output of BR is "new discoveries."
Applied research (AR) [p. 50]	Acquisition of new knowledge. It is, however, directed primarily toward a specific, practical aim or objective.	Applied research is undertaken either to determine possible uses for the findings of basic research or to determine new methods or ways of achieving specific and predetermined objectives.	The results of applied research are intended primarily to be valid for possible applications to products, operations, methods, or systems. Applied research gives operational form to ideas. The applications of the knowledge derived can be protected by intellectual property instruments, including secrecy. The main output of AR is an "invention" which can be protected through "patents."
Experimental development	Is intended to result in a plan or design for a new or substantially improved product or process, whether intended for sale or own use.	Based on past research or practical experience, it includes concept formulation, design, and the testing of product alternatives [p. 212]	It can include construction of prototypes and the operation of pilot plants [p. 212].

Source: Own elaboration based on the *Frascati Manual* of OECD [2015].

1.3 INNOVATION: DEFINITION AND TYPES

Resulting from the excessive use currently given to the concept of innovation, this term is mistakenly considered a recent business trend entirely linked to technological development. Perhaps, for this reason our interest goes beyond the mere definition of the term innovation, providing a more holistic view of the dimensions of the concept. Thus, we believe, it will be easier to understand the prominence and impact generated by the innovative fact.

The origin of the study of knowledge and the impact of changes and technological contributions to organizations date back to the 18th century, the result of the interest of economists to explain the economic development of capitalist society. Adam Smith and Karl Marx were ahead of other theorists in offering an explanation of the role that technology represents as a facilitator and generator of wealth.

We want to delve into the term innovation itself through a study of classical approaches to the theory of growth. Initially, we refer to Adam Smith, who, in his work *The Wealth of Nations*, first refers to technological progress, mentioning the role of machinery in the production process. Thus, he set his famous example of a pin maker, where he suggests that it is the division of labor itself as a way of producing that "probably" led to the "invention" of "the machinery employed in it." This view led to the endogenous growth theory approach, where savings in an economy lead to capital formation, which is a key factor for growth by means of labor productivity.

Additionally, Karl Marx, in his work *The Capital*, refers to the increase of technical productivity through specialization of the workforce and capital accumulation. Thus, a capitalist introduces a new and superior way of production to obtain a surplus of value, which is a motive to increase productiveness.

Another classic economist, David Ricardo, also links technological progress with economic evolution and growth. Hence, in his work *The Principles of Political Economy and Taxation*, he argues that the discovery and useful application of machinery always lead to an increase in the net production of the country, recognizing the importance of innovation as a way to enhance economic development.

The neoclassical approach led to a new concept, the production function, as a way to understand the relationship between the output of economic activity and the inputs needed. Under this approach, it is reasoned that the way inputs will be transformed into outputs will be determined by technology, which leads to the use of fewer inputs for the production output, increasing the productivity of the system itself. In this sense, technological change was depicted by movements over time of the production possibility frontier and, as a result, it was generally represented as a problem of maximization under constraints where its rate and direction derived from the rational choice of the representative firm [Swan, 1956, cited in Conte, 2006].

Since the mid-1980s, the so-called New Growth Theory or "endogenous growth theory" includes a two-fold update. First, it changes the view from previous theories that qualified technology as given to a product of economic activity, making technology a function of the market model. Likewise, this theory embraces knowledge and technology and characterized them by increasing returns, which leads to economic growth. "While exogenous technological change is ruled out, the model here can be viewed as an equilibrium model of endogenous technological change in which long-run growth is driven primarily by the accumulation of knowledge by forward-looking, profit-maximizing agents. This focus on knowledge as the basic form of capital suggests natural changes in the formulation of the standard aggregate growth model" [Romer, 1986, p. 1003].

Being aware of the different approaches to technological change in the economic literature, we do not lose our aim to offer an innovation definition. In order to do so, we consider that it is worth citing one of the seminal references in the innovation realm, from Joseph A. Schumpeter. This author, in his work *Business Cycles: A Theoretical, Historical, and Statistical Analysis of the Capitalist Process*, established the fundamentals of innovation. Schumpeter first made a distinction between the terms innovation and invention. In fact, he argues that both terms are not synonymous and have a distant relation between them, which can lead to misleading associations. Thus, he maintains that "innovation is possible without anything we should identify as invention and invention does not necessarily induce innovation, but produces of itself no economically relevant effect at all" [Schumpeter, 1939, p. 80]. Furthermore, he adds that even when innovation consists of putting into practice a particular invention, those two are entirely different things. According to Schumpeter's approach, an innovation necessarily entails the construction of a new plant and equipment, the setting up of a "New Firm," and the rise to leadership of "New Men" [Schumpeter, 1939]. This approach relates the concept of innovation to entrepreneurship, since every single innovation seems to be inserted in a new entrepreneurial venture. Thus, considering the issues raised by Schumpeter, we could say that innovation stems from the willingness, assertiveness, and wit of the leader in order to generate and promote the production of ideas and to involve the organization in their implementation through a process of innovation and change.

Schumpeter [1939, p. 84] argues that "innovation combines factors in a new way, or that it consists in carrying out New Combinations, those current adaptations of the coefficients of production which are part and parcel of the most ordinary run of economic routine within given production functions" and offered his renowned "trilogy" by differentiating between invention, innovation, and diffusion. He defends that, while the invention process includes the origination of new ideas and is commonly associated with science and basic research, the innovation process represents the development of new ideas into marketable products and processes, generally associated with technology and applied research, and determines the creation of economic value at the firm [Conte, 2006].

More precisely, Schumpeter defines innovation as a five-fold term that includes the following:

- the introduction of a new product or a qualitative change in an existing product;
- the introduction of a new method of production;
- the opening of a new market;
- the development of new sources of supply for raw materials or other inputs;
- the changes in industrial organization in a specific industry.

But, according to Schumpeter's theory, why do firms innovate and why does technical change happen? Technological changes are the foundation of some advantages for the innovator–entrepreneur. These technological changes lead to productivity increases, cost advantages over competitors, or market share gains through a monopolistic position in the market, which drive in rents for the innovator. Therefore, the reason put forward to explain innovation is the seeking of rents by innovators. When a firm establishes a monopolistic position in the market, it can be shown in higher prices and consequently in rent gains, and this will last as long as the company maintains this position. Yet, diffusion of new innovations leads competitors to implement new technological developments and to fill the gap between innovation leaders and followers throughout the adoption process.

Noting the differences between the concept of innovation and invention, we also have to cite the work of Freeman's *The Economics of Industrial Innovation* [Freeman, 1982, p. 7], which distinguishes between both terms, defining invention as an "idea, a sketch or a model for a product, process, or a new or improved system," while he depicts innovation in the economic sense, arguing that the term "refers only to the first commercial transaction of such product, process, or system..." [Freeman, 1982]. He also defines innovation as "the coupling of an inventive idea with a potential market" [Freeman, 1982, p. 289] or "the introduction and spread of new and improved products and processes in the economy and technological innovation to describe advances in knowledge [Freeman, 1982, p. 18]. At this point, we highlight that we are introducing two substantial elements within the concept of innovation — on the one hand the "generation" and on the other hand the "adoption." The first concerns the creation, development, initiation, or invention of something new, while the second is about the application, assimilation, deployment, transfer, and diffusion of the invented. This is important because part of the success of innovation and knowledge generation depends on the extent of the adoption and dissemination of them. Some authors like Afuah [2003] argue in favor of this "generation" approach, considering that innovation is the development of an idea or invention and its conversion into a useful application. Conversely, Roberts [2007] puts more focus on the adoption or dissemination process and describes innovation as a process of "invention plus exploitation"; i.e., there is a process of invention that creates a new idea and a process of exploitation, development, and commercial spread of innovation. Dismissing the concept of adoption qualifies innovation as just an invention or creation, bypassing the process of assimilation that should be undertaken by the organization or which is intended to assure the consumer himself.

Yet, the innovation concept requires a multifaceted approach. Thus, some authors like Dosi [1982] focus on the creative side of innovation, defining it as a problem-solving process. Other scholars pinpointed the interactive and multi-agent character of the innovation process [Kline and Rosenberg, 1986], while few others highlighted the importance of the knowledge exchange processes [Patel and Pavitt, 1994] or learning [Cohen and Levinthal, 1990] to define innovation.

Despite an abundance of innovation classifications and approaches, there are some common, basic typologies within the term innovation. The first typology of innovation we refer to is the distinction between radical or discontinuous innovation and incremental innovation. In this classification, we refer to the extent of the change that is being introduced and how it affects existing products, markets, or technologies. Radical innovation refers to "breakthroughs that change the nature of products and services and may contribute to technological revolutions" [Dodgson *et al.*, 2008, pp. 54–44]. Utterback also defines discontinuous change or radical innovation as the "change that sweeps away much of a firm's existing investment in technical skills and knowledge, designs, production technique, plant, and equipment" [Utterback, 1996, p. 200]. Conversely, incremental or continuous innovation includes minor changes in products, services, or processes that do not constitute a huge change and is mainly related to small improvements. Thus, Rothwell and Gardiner [1988, cited in García, 2002] refer to innovation and re-innovation, presenting a dichotomy where there is a distinction between incremental innovation (improving upon an existing product design), generational (improving existing products through new technology), improvements (improved materials in existing products), and minor details (new technology improving subsystems of existing products).

Unlike what happens with radical innovation, the role of R&D is not so relevant in incremental innovation, which is characterized by the introduction of gradual improvements coming from business experience, process reengineering, or improvements that have already been tested in other environments. Therefore, in other words, incremental innovation relates more to development than to research [García-Manjón and Escobar, 2010].

Connected to the concept of innovation radicalness is the degree of novelty of the innovation. Garcia [2002] refers to macro- and micro-approaches to explain the degree of novelty of a specific innovation. Thus, the macrolevel approach discriminates between what is new to the world, the market, or an industry, whereas the microlevel approach identifies if the product is new to the company or the client [Garcia, 2002, p. 118]. The degree of novelty is also linked to the concept of originality, depending on the use of new versus existing knowledge in order to produce innovations. Moreover, originality often implies more risk and uncertainty [Rosenberg, 1976].

Continuing our exploration of the innovation concept, we have to make reference to OECD [2018], which in its publication known as the *Oslo Manual*, offers a comprehensive definition of business innovation, outlining innovation as "a new or improved product or business process (or combination thereof) that differs significantly from the firm's previous products or business processes and that has been introduced on the market or brought into use by the firm" [p. 68]. Consequently, the *Oslo Manual* qualifies an innovating firm as one that "reports one or more innovations within the observation period. This applies equally to a firm that is individually or jointly responsible for an innovation" [p. 81].

So, the *Oslo Manual* distinguishes two types of innovation: "innovations that change the firm's products (product innovations), and innovations that change the firm's business processes (business process innovations)" [p. 70]. Then, we offer the following definitions from the *Oslo Manual*:

> Product innovation: "A product innovation is a new or improved good or service that differs significantly from the firm's previous goods or services and that has been introduced on the market" [p. 70]. This includes significant improvements "to one or more characteristics or performance specifications. This includes the addition of new functions, or improvements to existing functions or user utility. Relevant functional characteristics include quality, technical specifications, reliability, durability,

economic efficiency during use, affordability, convenience, usability, and user friendliness." [p. 71]

Business process innovation: "A business process innovation is a new or improved business process for one or more business functions that differs significantly from the firm's previous business processes and that has been brought into use in the firm" [p. 72]. "A business process innovation can involve improvements to one or more aspects of a single business function or to combinations of different business functions. They can involve the adoption by the firm of new or improved business services that are delivered by external contractors, for instance accounting or human resources systems" [p. 72].

REFERENCES

Afuah, A. [2003]. *Innovation Management: Strategies, Implementation, and Profits*, 2nd edn. (OUP, New York, USA).

Aschhoff, B., Crass, D., Cremers, K., Grimpe, C., Rammer, C., Brandes, F., and Montalvo, C. [2010]. *European Competitiveness in Key Enabling Technologies: Summary Report*. European Commission.

Biagi, F. [2013]. ICT and productivity: A review of the literature. JRC Institute for Prospective Technological Studies, *Digital Economy Working Paper*, No. 9.

Botsman, R. [2015]. Defining the sharing economy: What is collaborative consumption–and what isn't. *Fastcoexist.com*, p. 27.

Bresnahan, T. F., and Trajtenberg, M. [1995]. General purpose technologies "Engines of growth?" *Journal of Econometrics*, 65(1), 83–108.

Brooks, H. [1994]. The relationship between science and technology, *Research Policy*, 23(5), 477–486.

Bunge, M. [1966]. Technology as applied science, Rapp, F. (ed.), *Contributions to a Philosophy of Technology* (Springer, Dordrecht) pp. 19–39.

Bush, V. [1945]. *Science — The Endless Frontier: A Report to the President on a Program for Postwar Scientific Research* (National Science Foundation, Washington).

Chesbrough, H. W. [2003]. The era of open innovation, *MIT Sloan Management Review*, 44(3), pp. 35–41.

Cohen, W. M., and Levinthal, D. A. [1990]. Absorptive capacity: A new perspective on learning and innovation. *Administrative Science Quarterly*, 35, pp. 147–160.

Conte, A. [2006]. The evolution of the literature on technological change over time: A survey, *Papers on Entrepreneurship, Growth and Public Policy*, No. 0107.

Dodgson, M., Gann, D. M., and Salter, A. [2008]. *The Management of Technological Innovation: Strategy and Practice*, 2nd edn. (Oxford University Press, New York).

Dosi, G. [1982]. Technological paradigms and technological trajectories: A suggested interpretation of the determinants and directions of technical change, *Research Policy*, 11(3), pp. 147–162.

Freeman, C. [1982]. *Economics of Industrial Innovation* (MIT Press, USA).

García, R. [2002]. A critical look at technological innovation typology and innovativeness terminology: A literature review, *Journal of Product Innovation Management*, 19(2), pp. 110–132.

Giere, R. N. [1993]. Science and technology studies: Prospects for an enlightened postmodern synthesis, *Science, Technology, & Human Values*, 18(1), pp. 102–112.

Goldin, C., and Katz, L. F. [1998]. The origins of technology-skill complementarity, *The Quarterly Journal of Economics*, 113(3), pp. 693–732.

Gordon, R. J. [2012]. *Is US Economic Growth Over? Faltering Innovation Confronts the Six Headwinds*, No. w18315, National Bureau of Economic Research, USA.

Gordon, R. J. [2000]. Does the "new economy" measure up to the great inventions of the past? *Journal of Economic Perspectives*, 14(4), pp. 49–74.

Grübler, A. [2003]. *Technology and Global Change* (Cambridge University Press).

Hanna, N. K. [2010]. *Transforming Government and Building the Information Society: Challenges and Opportunities for the Developing World* (Springer Science & Business Media, USA).

Jorgenson, D. W., Ho, M. S., and Stiroh, K. J. [2008]. A retrospective look at the U.S. productivity growth resurgence, *Journal of Economic Perspectives*, 22(1), pp. 3–24.

Jovanovic, B., and Rousseau, P. L. [2005]. General purpose technologies, Aghion, F., and Durlauf, S. (eds.), Chapter 18, Handbook of Economic Growth (Elsevier, Holland), pp. 1181–1224.

Kline, S., and Rosenberg, N. [1986]. An overview of innovation, Landau, R., and Rosenberg, N. (eds.), *The Positive Sum Strategy: Harnessing Technology for Economic Growth* (National Academy of Sciences, Washington) pp. 275–306.

Knell, M. [2010]. Nanotechnology and the sixth technological revolution, Cozzens, S. E., and Wetmore, J. M. (eds.), *Nanotechnology and the Challenges of Equity, Equality and Development* (Springer, Dordrecht) pp. 127–143.

Kondratieff, N. D. [1935]. The long waves in economic life, *The Review of Economics and Statistics*, 17(6), pp. 105–115.

Kuhn, T. S. [1962]. *The Structure of Scientific Revolutions,* 1st edn. (University of Chicago Press, Chicago).

Kurzweil, R. [2006]. *The Singularity Is Near: When Humans Transcend Biology* (Penguin Group, New York).

Layton, E. [1971]. Mirror-image twins: The communities of science and technology in 19th-century America, *Technology and Culture*, 12(4), p. 562.

Lipsey, R., Bekar, C., and Carlaw, K. [1998]. *General Purpose Technologies and Economic Growth*, Helpman, E. (ed.), Chapter 2, *What Requires Explanation*? (MIT Press, Cambridge, MA) pp. 15–54.

Manjón, J. V. G., and Escobar, J. A. R. [2010]. *El ABC de la innovación: Principales definiciones, modelos y concepto*s (Netbiblo, Spain).

OECD [2008]. *Open Innovation in Global Networks* (OECD Publishing, Paris).

OECD [2010a]. *Improving Health Sector Efficiency* (OECD Publishing, Paris).

OECD [2010b]. *The OECD Innovation Strategy Getting a Head Start on Tomorrow* (OECD Publishing, Paris).

OECD [2015]. F*rascati Manual* 2015: *Guidelines for Collecting and Reporting Data on Research and Experimental Development*; The Measurement of Scientific, Technological and Innovation Activities (OECD Publishing, Paris).

OECD [2018]. *Oslo Manual* (OECD Publishing, Paris).

Patel, P., and Pavitt, K. [1994]. National innovation systems: Why they are important, and how they might be measured and compared, *Economics of Innovation and New Technology*, 3(1), pp. 77–95.

Pérez, C. [2003]. *Technological Revolutions and Financial Capital* (Edward Elgar Publishing).

Pérez, C. [2004]. Finance and technical change: A long-term view, Pyka, A. H. (eds.), *The Elgar Companion to Neo-Schumpeterian Economics* (Edward Elgar, Cheltenham) pp. 775–799.

Pérez, C. [2016]. *Rethinking Capitalism: Economics and Policy for Sustainable and Inclusive Growth*, Jacobs, M., and Mazzucato, M. (eds.), Chapter 11, *Capitalism, Technology and a Green Global Golden Age: The Role of History in Helping to Shape the Future* (Wiley, USA) pp. 191–217.

Pitt, J. C. [1995]. *New Directions in the Philosophy of Technology* (Springer Science & Business Media).

Powell, W. W., and Snellman, K. [2004]. The knowledge economy, *Annual Review of Sociology*, 30, pp. 199–220.

Price, D. P., Stoica, M., and Boncella, R. J. [2013]. The relationship between innovation, knowledge, and performance in family and non-family firms: An analysis of SMEs, *Journal of Innovation and Entrepreneurship*, 2(1), p. 14.

Resnick, M. [2002]. *Rethinking Learning in the Digital Age. The Global Information Technology Report: Readiness for the Networked World* (Oxford University Press, USA).

Roberts, E. B. [2007]. Managing invention and innovation. *Research-Technology Management*, 50(1), pp. 35–54.

Roco, M. C., and Bainbridge, W. S. (eds.), [2003]. *Converging Technologies for Improving Human Performance* (Springer, Netherlands).

Romer, P. M. [1986]. Increasing returns and long-run growth, *Journal of Political Economy*, 94(5), pp. 1002–1037.

Rosenberg, N. [1976]. *Perspectives on Technology* (CUP Archive).

Rothwell, R., and Gardiner, P. [1988]. Re-innovation and robust designs: Producer and user benefits, *ResearchGate*, 3(3), pp. 372–387.

Satell, G. [2016]. Reasons to Believe the Singularity is Near, *Forbes*.

Schumpeter, J. A. [1939]. *Business Cycles. A Theoretical, Historical and Statistical Analysis of the Capitalist Process*, 2 vols (McGraw-Hill, New York).

Selinger, M. [2004]. Cultural and pedagogical implications of a global e-learning programme, *Cambridge Journal of Education*, 34(2), pp. 223–239.

Smith, K. [2000]. Innovation as a systemic phenomenon: Rethinking the role of policy. *Enterprise and Innovation Management Studies*, 1(1), pp. 73–102.

Solow, R. [1987]. We'd Better Watch Out. *New York Times* Book Review.

Swan, T. W. [1956]. Economic growth and capital accumulation. *Economic Record*, 32(2), pp. 334–361.

Trajtenberg, M., and Bresnahan, T. F. [1992]. General purpose technologies: "Engines of growth?" *NBER Working Paper*, No. 4148.

Utterback, J. M. [1996]. *Mastering the Dynamics of Innovation*, 2nd edn. (Harvard Business Review Press, Boston).

van Ark, B., O'Mahony, M., and Timmer, M. P. [2008]. The productivity gap between Europe and the United States: Trends and causes. *Journal of Economic Perspectives*, 22(1), pp. 25–44.

CHAPTER 2

INNOVATION AS A PROCESS

2.1 THE PROCESS AND MANAGEMENT OF INNOVATION

Both scholars and practitioners have sought the measurement of innovation success in order to understand the different processes that comprise it and how to intervene over them. The interest in analyzing and managing the innovation process itself is linked with the aim of achieving higher rates of success and improving innovative output, and at the same time lowering the high risk of uncertainty inherent in the innovation process itself. The risky and uncertain character of innovation practices has led some authors to refer to it as the "innovation journey" [Van de Ven, 1999]. Therefore, due to its uncertainty, it has been in the interest of academics and practitioners to know how the innovation process can be managed. In the same vein, Kline and Rosenberg [1986, pp. 275–276] point out that "Since innovation, by definition, involves the creation and marketing of the new, these gauntlets, singly and in combination, make the outcome of innovation a highly uncertain process. Thus, an important and useful way to consider the process of innovation is as an exercise in the management and reduction of uncertainty."

The conceptualization of innovation as a process is implicit in its own nature and varies depending on the approach taken by different authors. Thus, Dosi [1982] refers to the innovation concept as a problem-solving process. Kline and Rosenberg outline innovation as an interactive and multi-agent process underlying the systemic character of the innovation process and discriminating between the optimization and transformational approaches. Thus, the authors postulated that, "In the contemporary economy, some portion of business behavior is closely calculated by sophisticated optimization methods. Another portion is innovation activity shaped by the creative problem-solving insights of scientists, engineers, and managers" [Kline and Rosenberg, 1986, p. 135]. This dichotomy between transformation and execution is an interesting approach that is going to lead the behavior of managers and the way business is organized. While the transformational approach demands more flexible and open processes, the optimization realm requests more rigid processes and organizational methodologies.

Another motivating and well-known approach is the definition of innovation as a learning process. Being aware of the importance of knowledge as the main corporate resource, the business focuses its efforts on determining different knowledge flows in order to manage them. Some scholars have also underlined the role of basic research in firm learning as a way to generate the most valuable resources of the firm [Cohen and Levinthal, 1990; Nonaka and Takeuchi, 1995; Cross and Israelit, 2009, p. 62]. Thus, the authors stated, "Emphasizing the role of basic research in firm learning, our perspective redirects attention from what happens to the knowledge outputs from the innovation process to the nature of the knowledge inputs themselves. Considering that absorptive capacity tends to be specific to a field or knowledge domain means that the type of knowledge that the firm believes it may have to exploit will affect the sort of research the firm conducts." In accordance with the latter definition, Edquist [1997] also defines innovation as an interactive process of learning and exchange where interdependence between actors generates an innovative system or an innovation cluster, stressing the systemic character of the innovation process. Usually, the innovation process begins with an idea, followed by several stages and culminating with the introduction into the market of new products, the implementation of new processes, or early termination due to the failure of the innovation process. For a better understanding of this process, scholars and practitioners have divided it into different stages that are visible milestones in innovative development. Throughout the literature review in the area, it is clear that the innovation process is composed of a series of basic activities that ensure the innovative output of the firm, the generation of new products, or the implementation of new production or distribution processes. These aforementioned activities encompass the identification of needs and idea generation [Utterback, 1971], the development of new products [Tidd and Hull, 2003; Kahn, 2012; Cooper, 2011; Wheelwright, 2011], the discovery and generation of new knowledge [Nonaka and Takeuchi, 1995; Usher, 2011; Cohen and Levinthal, 1990], the creation of prototypes and models, the assimilation and implementation of new technology [Saad, 2000; Bennett and Kerr, 1996], the implementation of new production and distribution processes [Davenport, 1995; Pisano, 1996], the implementation of new marketing, strategy, and management activities [Burgelman *et al.*, 2008], and the learning process involved in the innovation practice [Leonard-Barton, 1995; Cohen and Levinthal, 1990; Ayas, 1997]. Consequently, it leads authors such as Lendel *et al.* [2015, p. 862] to outline innovation as a "sequence of activities aimed at creation and implementation of innovation. It includes activities related to generating innovative ideas, their evaluation, creation of innovation, and ensuring its spreading among customers."

However, the innovation process is far too complex to be well-defined or depicted as a step-by-step procedure. Some authors consider that the first stage in the innovation process is recognizing the need to innovate. It then becomes innovation generation, and finally, diffusion, adoption, and use of the innovation [Knight, 1967; Bessant and Tidd, 2015]. Yet, some authors argue that "innovation is more than simply coming up with good ideas; it is the process of growing them into practical use" [Tidd and Bessant, 2013, p. 64], enhancing the importance of the market (diffusion and adoption) stemming from the Schumpeterian view.

Other contributions put in doubt the continuous character of the innovation process, describing it as random phenomena. This is the case of Usher [2011, p. 6] who states, "The process of innovation has frequently been held to be an unusual and mysterious phenomenon of our mental life. It has been long regarded as the result of special processes of inspiration that are experienced only by persons of the special grade called men of genius" [Usher, 2011, p. 8]. This "magic" character of the innovation stream makes it difficult to view innovation as a process that can be divided into different stages and consequently be organized and managed.

Conversely, it is also worth citing the vision of Utterback, who considers that the innovation process is overlapped into three subprocesses, namely idea generation, problem solving, and implementation of the innovation. Possibly, the first two processes conclude in an invention and the last one results in an innovation. The idea generation phase, argues the author, "results in origination of a design concept or technical proposal, perhaps via synthesis of several pieces of existing information. The problem-solving phase results in an original technical solution, or an invention. The implementation phase results in market introduction of the original solution making it an innovation as defined above. Diffusion is the mechanism of communication and increasing use through which an innovation comes to have a significant economic impact. It is not strictly a part of the process of innovation as defined, as it occurs in the firm's outside environment" [Utterback, 1971, p. 78].

White and Bruton [2010] argue that no matter how you consider the innovation process, what becomes clear is that innovation should be a continuous process in the organizational realm. The authors point out that innovation is not an isolated event that occurs once, bringing all the needed innovation into the firm, but a cyclical and continuing phenomenon. The innovation process comprises different phases or subprocesses: forecast changes in technology that include feasibility studies about the scientific and technical market, acquisition of technology through transfer mechanisms, the integration and exploitation of technology, and finally the implementation of technology incorporating design, testing, and roll-out [White and Bruton, 2010, p. 22]. This approach stresses the importance of technology and the diffusion and adoption process as a way to develop innovation.

Tidd and Bessant included the process approach in the definition of innovation itself. Thus, these authors describe innovation as, "the core process within an organization associated with renewal, with refreshing what it offers the world and how it creates and delivers that offering." The authors argue "that considering innovation as the only way to face the future of a company, since they consider innovation to be a generic activity associated with survival and growth. And at this level of abstraction we can see the underlying process as common to all firms" [Tidd and Bessant, 2013, p. 67]. Therefore, these authors identify innovation as a strategic process within the firm that is directly related to the value proposal to the market, underlying the continuous character of innovation as a way to refresh the firm's value proposition.

The authors propose that the innovation process contains four main phases, which are searching, selecting, implementing, and learning. More specifically, they describe each of the phases. The "searching process" is aimed at finding relevant signals about threats and opportunities by scanning the environment. The "selection process" analyzes and interprets these signals in order to decide which ones the firm can better focus on and responds to them. The "implementation process" is the central one, and it deals with the process of converting these ideas into "something new" internally and to the market. This implementation stage entails the acquisition of knowledge to enable innovation, innovation execution, innovation management, and definitively sustaining adoption. Finally, the authors stress the "learning process" of building a knowledge base and learning from the process management itself.

Another noteworthy approach comes from Birkinshaw and Hansen [2007], who in their article *The Innovation Value Chain*, present innovation as a sequential, three-phase process that involves idea generation, idea development, and the diffusion of developed concepts. According to the authors, in order to improve innovation, companies need to see innovation as the process of transforming ideas into marketable products, similar to the Porter value-added approach in which raw materials were transformed into finished goods. Thus, the first phase in the chain is "idea generation" which can occur inside a unit, across units in a company, or outside the firm. The second phase is the conversion of ideas into products or practices through a selection process. Finally, the third phase is the diffusion of those products and practices internally and externally. This approach does not differ too much from the former schemes that have been presented. As for the way the innovation process is conducted, we can cite Hidalgo and Albors [2008] who say that the innovation process can be internally and externally driven. Internally, innovation is influenced by senior management attitudes, marketing, organization development, or IT departments. The authors highlight that collaborative efforts support and facilitate the innovation process. These efforts can be evidenced in senior management devoting time to investigate the future needs of the marketplace, working environments enhancing creative solutions, support for joint ventures, and good project management. From an external perspective, innovation management is driven by different knowledge-intensive organizations that build knowledge as their primary value-adding process.

Tidd and Bessant [2013] argue that the importance of understanding the innovation process is related to the capacity to administer it, since, although it is difficult to find someone who does not recognize the importance of the innovation process, it is also true that there is an assumption of the difficulty to manage it. The authors add, "But despite the uncertain and apparently random nature of the innovation process, it is possible to find an underlying pattern of success. Not every innovation fails, and some firms (and individuals) appear to have learned ways of responding and managing it such that, while there is never a cast-iron guarantee, at least the odds in favor of successful innovation can be improved" [Tidd and Bessant, 2013, p. 80].

Once we assume that innovation is a process that includes all the activities that lead the company to generate innovative outputs, we can define the management of innovation as the conscious process of organization, control, and execution of all the activities that compose it. In this way, we are interested in shedding light on how to positively influence the innovation process; i.e., what parameters and factors are put into practice by those organizations that demonstrate better innovative performance. If we identify and understand the activities and conditions that trigger organizational innovation, we could manage those activities to achieve our innovative goals. To fulfil this objective, some authors have studied successful companies and describe how they organize and manage innovation [Van de Ven and Poole, 1990; Rothwell *et al.*, 1974]. Accordingly, Birkinshaw and Hansen [2007] argue that among all the phases of the innovation chain, there are six that we can classify as critical, since they directly affect the success of the innovative process. These tasks are internal sourcing (number of high-quality ideas generated within the unit), cross-unit sourcing (cross-pollination or number of high-quality ideas generated across business units), external sourcing (collaboration with partners outside the firm), selection (percentage of ideas that end up being selected and funded), development (percentage of generated ideas that lead to revenues), and companywide spread of the idea (penetration into desired markets and customers). The authors assume that while some companies excel in some of them, conversely there are others struggling with some aspects.

In the same vein, Van der Panne *et al.* [2003] examined up to 43 investigations on the factors of success in innovation. Subsequently, they found consistent data on the relevance of factors such as firm culture, experience with innovation, the multidisciplinary character of the R&D team, and explicit recognition of the collective character of the innovation process or the advantages of the matrix organization. However, there are factors that remain inconsistent or inconclusive, among which we can cite strength of competition, R&D intensity, the degree to which a project is "innovative" or "technologically advanced," and top management support. Likewise, Smith *et al.* [2008, p. 6] studied the factors influencing the organization's ability to manage innovation. The authors analyzed 102 investigations through a structured literature review. As a result, they established a series of factors and subfactors that influence innovation: technology (utilization of technology, technical skills and education, technology strategy), innovation process (idea generation, selection and evaluation techniques, implementation mechanism), corporate strategy (organizational strategy, innovation strategy, vision and goals of the organization, strategic decision-making), organizational structure (organizational differentiation, centralization, formality), organizational culture (communication, collaboration, attitude toward risk, attitude toward innovation), employees (motivation to innovate, employee skills and education, employee personalities, training), resources (utilization of slack resources, planning and management of resources, knowledge resources, technology resources, financial resources), knowledge management (organizational learning, knowledge of external environment, utilization of knowledge repositories), and management style and leadership (management personalities, management style, motivation of employees).

Many authors prioritize the importance given to corporate culture in order to trigger innovation, since they recognize the pervasive role that organizational culture plays in the management of innovation. Thus, Ahmed [1998] categorizes it as the primary determinant of innovation. Although the scientific literature maintains an agreement on the importance of corporate culture in innovation, it is more difficult to find consensus on how organizational innovations influence the effective management of innovation. Thus, Mintzberg [1979, p. 209] describes the "innovative organization" by simply describing an optimal organizational structure for innovation management. According to Mintzberg, "Every one of its characteristics is very much in vogue today: emphasis on expertise, organic structure, project teams, task forces, decentralization of power, matrix structure, sophisticated technical systems, automation, and young organizations. Thus, if professional and diversified forms are yesterday's configurations, and the entrepreneurial and machine forms yet earlier configurations, then the innovative form is clearly today's." Equally, Burns and Stalker [1961] facilitate a contingent approach that shows the complexities of the organizational structure in the management of innovation. Different authors agree that the abandonment of autocratic systems toward participatory and more democratic systems facilitates innovation [Roffe, 1999; Rivas and Gobeli, 2005; Hyland and Beckett, 2005]. The effects of organizational structure on innovation are more profoundly approached in Chapter 5.

As we stated before, learning is an important organizational process which is linked with the generation, management, and application of knowledge as the most valuable organizational asset. Therefore, existing literature seems to show consensus in the prominent role of knowledge management and learning in organizations. In this way, organizations that actively manage their knowledge and learning significantly increase their innovative performance [Pavitt, 2002; Jones, 2000]. Consequently, innovativeness reflects the extent to which new knowledge is embedded in new products or processes [Dewar and Dutton, 1986]. Likewise, an emphasis on organizational learning and development enhances the production of new ideas [Damanpour, 1987; Hurley and Hult, 1998], creativity, and problem-solving improvement [Senge, 1990; King and Anderson, 1990]. Technology is another component to take into account as an element that positively influences knowledge management and innovative performance. In particular, ICTs impact employee knowledge management, facilitating knowledge transfer [Sørensen and Stuart, 2000; Kandampully, 2002]. They can also generate joint repositories of knowledge within the organization [Damanpour, 1987; Jantunen, 2005].

Continuing with our review, the autonomy of the worker is another element that positively influences innovation in organizations. Apparently, workers who have a greater degree of control over their work feel comfortable in their innovative role in the organization [Tang, 1998; Zwetsloot, 2001; Amar, 2004]. However, there are also authors who point out the importance of supervisors' support to workers, since it facilitates their ability to innovate [Knight, 1987; Martins and Terblanche, 2003]. Corporate culture and leadership and their influence on innovation will be discussed in detail in Chapter 6.

Technology is a key factor in enhancing innovativeness, and it is embedded throughout the phases of the innovative process [Petroni, 1998; Loewe and Dominiquini, 2006], from idea generation to the arrival of new products or services to the market or the implementation of new processes. The literature point outs that employees are an interesting and rich source of ideas, and they should be stimulated to participate in the early stages to guarantee a continuous supply of ideas to the innovation process [Woodman *et al.*, 1993; Guimaraes and Langley, 1994; Andriopoulos and Lowe, 2000; McAdam and McClelland, 2002]. Some authors do insist on the need to train workers so that their participation has a positive impact on the innovation process [Koen and Kohli, 1998; Pohlmann *et al.*, 2005; Brennan and Dooley, 2005; Shipton *et al.*, 2006].

It is also worth citing the view posed in the *Oslo Manual* [OECD and Eurostat, 2005, p. 22] that highlights the importance of understanding what characteristics make firms more or less innovative and how innovation is generated within businesses. Thus, the *Oslo Manual* upholds that the propensity of a firm to innovate depends, of course, on the technological opportunities it faces. But it adds that, in addition, "Firms differ in their ability to recognize and exploit technological opportunities. In order to innovate, a firm must figure out what these opportunities are, set up a relevant strategy, and have the capabilities to transform these inputs into a real innovation, and do so faster than its competitors. But it would be misleading to stop there."

The *Oslo Manual* also highlights the importance of some factors such as skilled employees as key assets for an innovative firm. "Without skilled workers, a firm cannot master new technologies, let alone innovate. Apart from researchers, it needs engineers who can manage manufacturing operations, salespeople able to understand the technology they are selling (both to sell it and to bring back customers' suggestions), and general managers aware of technological issues." But capabilities for innovation also depend on the characteristics of the firm. Thus, the *Oslo Manual* points out that the structure of its labor force and facilities (skills, departments), its financial structure, its strategy on markets, competitors, alliances with other firms or with universities, and above all its internal organization are all key points for innovation. Therefore, it is necessary to focus on these elements in order to manage the innovation process adequately. Once we have analyzed innovation as a series of processes that seek to obtain business results under uncertainty and risk, it is useful to delve into the different models that try to explain innovation within organizations. In this way, we will be able to take decisions to manage them, so that we can focus our efforts on strengthening those elements of the process that have a more positive influence on the organization's innovative performance.

2.2 INNOVATION MODELS

One of the main challenges faced by authors studying the realm of technological innovation is understanding innovative behavior within organizations. The way an organization innovates has been depicted through the establishment of different models that try to sketch the innovation process itself. We have to highlight the difficulty for the establishment of these models, since innovation does not follow predefined guidelines. Accordingly, Saren [1984] recognizes the complexity of the innovation process, which challenges its management, explanation, and study and also presents problems for the development of a generalized model.

Many authors have studied the diverse innovation models present in academic literature in the past [Saren, 1984; Forrest, 1991; Rothwell, 1994; Niosi, 1999; Verloop, 2004]. The existence of innovation models implies a system approach where "everything interacts with everything but recognizes that in practice, some interactions matter more than others" [Padmore *et al.*, 1998, p. 605]. In Table 2.1, we provide a list of the main contributors and an outline of their proposals.

All the classifications follow a common structure, yet one of the best known innovation model classifications was developed by Rothwell [1994]. The author presents an innovation modeling framework composed of five different models. In what follows, we present an overview of Rothwell's classification.

Table 2.1. Innovation Models.

Author	Innovation models
Saren [1984]	The author reviewed the literature and posed a taxonomy of innovation models, which include five types: departmental-stage models, activity-stage models, decision-stage models, conversion process models, and response models.
Forrest [1991]	According to the author, there are four different innovation models: stage models, conversion models, technology-push and market-pull models, integrative models, and decision models.
Rothwell [1994]	A framework of five different generations of innovation models is depicted by this author.
Padmore *et al.* [1998]	The authors distinguished linear, chain-link, and cycle models. Linear models are split into stages: discover, develop, and deploy. The reference for the chain-link model is that of Kline and Rosenberg [1986]. The cycle models are based on Porter's value chain.
Trott [2008]	The author presents a classification comprising serendipity models, linear models, simultaneous coupling models, and interactive models.

Source: Own elaboration based on the literature.

2.2.1 *First Generation of Innovation Models (1950s to Mid-1960s)*

We can start talking about the first generation of innovation models, the so-called technology-push or science-push models, which treated innovation as a sequential process that took place in discrete stages. These models emerged in the 1950s and remained in force until the mid-1960s, placing them in the post-war period characterized by production problems and demand exceeding offering of products and services. Rothwell referred to the emergence of new industries that based their development on new technological opportunities; e.g., semiconductors, pharmaceuticals, electronic computing, and synthetic and composite materials. At the same time, there was the technology-led regeneration of existing sectors such as textiles and steel and the rapid application of technology to enhance the productivity and quality of agricultural production.

The origin of these models stems from Schumpeter's innovation theory and is directly related to the role of the entrepreneur from a double perspective: in the first stage embracing the vision of the entrepreneur,

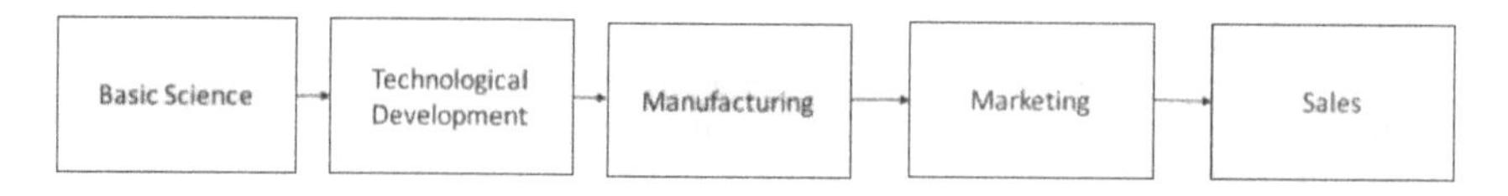

Figure 2.1. Technology-Push Models.
Source: Rothwell [1994].

founder of the company, and acting individually — action that receives the denomination of "heroic" by Schumpeter — to a later vision in which Schumpeter poses a bureaucratic model where innovation becomes a professional routine for R&D departments. These models can be summarized in Fig. 2.1.

Linear models have been mentioned in the literature as "assembly-line models," "pipeline models," "ladder models," and the "bucket models." In these models, innovation is understood as the progress from basic scientific knowledge, development, production, and marketing activities as a way of placing new products in the market. Linear models assumed that scientific discovery preceded and pushed technological innovation through applied research, engineering, manufacturing, and marketing strategies [Hobday, 2005]. Linear models also explain the production of knowledge as an output produced in a certain phase and sequentially transferred to the next level as inputs [Verloop, 2004].

Padmore *et al.* [1998, p. 607] distinguish three stages from discover to develop and deploy. The authors claim that in linear models, everything seems to originate with the discovery and "science has to be completed and packaged before becoming available for an invention." Innovation is understood as a linear process that has its origin in research and therefore in the generation of new knowledge, avoiding, at least theoretically, the possibility that the innovative dynamics could start from any other stages of the process. Padmore's vision was often used to justify R&D expenditures by firms and governments, as it was argued that this would lead to more innovation and, in turn, to faster economic growth. Research is the starting point and the basis for all innovations. It implies that the path from research through development and production is a standard and major pathway of innovation in both companies and economies. However, there is no feedback loop into the system. The attitudes in society were favorable to support scientific advances and industrial innovation, since both were seen as a way to overcome societal problems. Rothwell summarizes the vision for this model assuming that "more R&D in" results in "more successful new products out," assuming a direct relationship between research and development led to the generation of knowledge and the innovative performance obtained. According to Branscomb [1993], this kind of model overestimates the meaning of novelty — of "newness to universe" — as the singular foundation of the economic and social benefit resulting from R&D or scientific technological activity.

This model has been criticized because it offers a simplistic vision of the problem of innovation; hence, the OECD and Eurostat [2005] through the *Oslo Manual* recognize two fundamental shortcomings. On the one hand, too much emphasis is placed on research and technological development in confrontation with a broader view of innovation adopted by this body. On the other hand, it criticized the absence of feedback routes of the model within the development process or the market itself.

Other authors, such as Padmore *et al.* [1998, p. 607] argue that the model has been criticized because it does not seem to reflect what scientists, inventors, and innovators do and because it seems too simple. He affirms that "the linear model suggests that everything originates with the discovery" and adds "the invention is then perfected before being applied."

However, we must also assume that these models have served to outline the complex dynamics of innovation in such a way that, through them, we are able to unravel the various stages necessary in the innovative process.

2.2.2 *The Second-Generation Innovation Process (Mid-1960s to Early 1970s)*

Empirical studies of innovation processes began to emphasize market-led theories of innovation. Thus, linear models evolve over time, giving a greater value to the market's weight in the innovation process, introducing "demand-pull" or "market-pull" models. These models have been developed from the late 1960s to the early 1970s, coinciding with an intensified offer and rivalry. While manufacturing output and prosperity kept growing, the employment was static due to the growth in manufacturing productivity. Production was mainly based on existing technologies, and

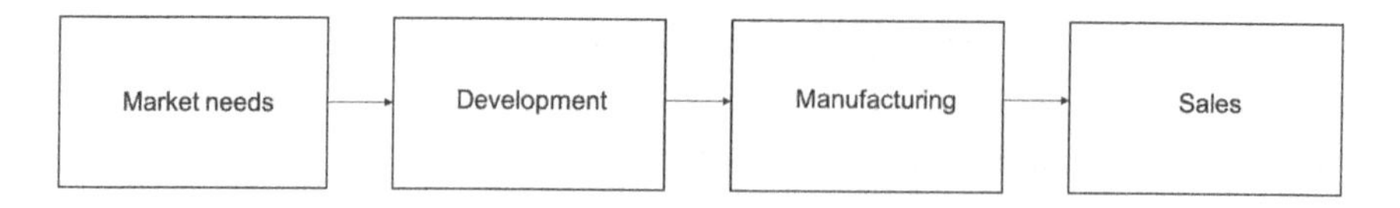

Figure 2.2. Market-pull Models.
Source: Rothwell [1994].

supply and demand were balanced. Thus, expansionary technological change was substituted by the rationalization of technology and a greater market weighting. While we enjoyed sufficient demand to absorb all the existing production, we faced the challenge of adapting supply to an increasingly demanding market, with the market being a source of innovations. These models stress the role of the marketplace and market research in recognizing and responding to customer needs, as well as focusing R&D investments on those needs. Even public policies adopted this approach, using public procurement to support certain industrial policies. In Fig. 2.2, we can see a representation of this type of model according to Rothwell.

This typology of models was also preferred by authors such as Schmookler [1962], highlighting the importance of the market in the innovation process. Consumer needs are the key point in these models, so the market and consumer needs are a source of ideas for R&D departments. As in previous models, the weakness of the model lies in the lack of references to scientific and technological knowledge, which is essential for innovation.

Hayes and Abernathy [1980] recognized that one of the weakest points inherent in this model was that it could lead companies to neglect long-term R&D programs and become locked into a regime of technological incrementalism as they adapted existing product groups to meet changing user requirements along maturing performance trajectories.

Another approach under the linear model is the departmental-stage model. These models, like the previous ones, consider innovation a sequential activity. The innovation process is considered a series of consecutive stages, detailing and emphasizing the particular activities that take place either in each of the stages or in the departments involved. One of the main contributions of this model is that they include elements from both approaches: the technology-push and the pull of demand. In its simplest form, the process consisted of two stages, the conception of an idea or an invention, followed by a second stage that led to the commercialization of this idea. This approach is followed by Utterback who describes the process of innovation in simple terms but adds one more phase to arrive at a three-stage model that includes idea generation, implementation, and dissemination [Utterback, 1971; Abernathy and Utterback, 1978; Utterback and Abernathy, 1975]. Following Mansfield [1968], the processes range from invention to innovation and diffusion. The different stages in Mansfield's model are entirely referred inside the firm, with no concern to the environment or society at large.

Finally, authors such as Saren [1984] describe the process of innovation in terms of the departments involved, where an idea becomes an input for the R&D department, from there to design, engineering, production, marketing, and finally, as output of the process, the product. One of the weaknesses of these models is that they consider each activity or department as individual and isolated from the rest, when there are numerous interrelationships [Forrest, 1991].

Assuming the great deficiencies of linear models that force their evolution, these lead us to the next generation of models that are the so-called interactive models or coupling models. These models introduce the concept that innovation takes place within an environment with which the company interacts.

2.2.3 *The Third-Generation Innovation Process (Early 1970s to Mid-1980s)*

This period was characterized by high rates of inflation and demand saturation that led to a situation of supply exceeding demand which in turn led to the rise of unemployment. This situation derives from technologically consolidated strategies and the search for successful innovation. Diverse empirical studies during the 1970s revealed that previous linear models hardly explained the industrial innovation processes, since the process neglected the interaction between different agents and a complex relationship with the environment. Thus, some authors such as

Mowery and Rosenberg [1979] argued that innovation was portrayed by an interaction or coupling between science and technology (S&T) and the market, with technological capabilities and market needs overcoming the pitfalls of the previous models. More specifically, and according to Rothwell and Zegveld [1985, p. 50], innovation is "a complex network of communication paths, inside and outside the organization, related to internal functions and to the scientific and technological community and the market itself." In this way, a broader vision of the innovative process is offered, not only as a sequential organization of different activities but also considering the existence of internal and external elements of the organization, which also has its importance in the innovative process of the organization. Rothwell also observed that the interaction process was not necessarily continuous but could be explained in terms of interactive and interdependent functional stages involving complex interaction paths and intra- and inter-organizational linkages. Unlike the two previous models, the interactive model explicitly links decision-making with the scientific and technology community and with the market.

Next, we can see in Fig. 2.3 a representation of the interactive model of innovation proposed by Rothwell and Zegveld.

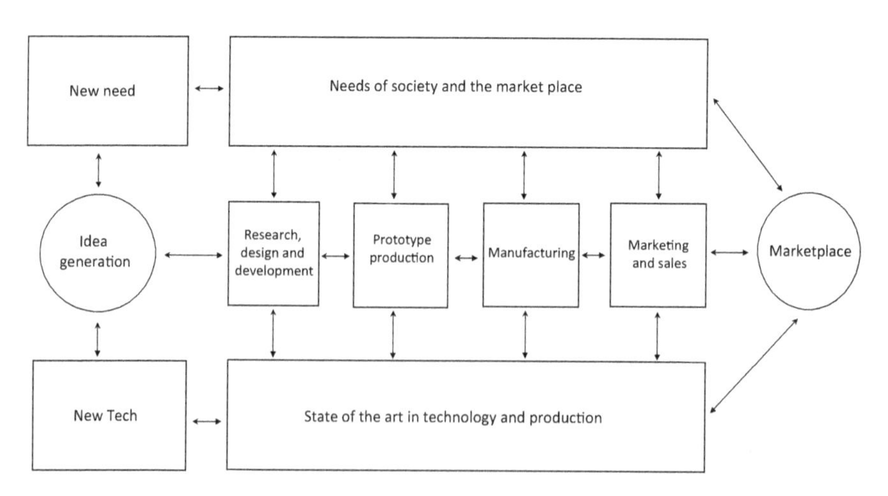

Figure 2.3. The Coupling or Interactive Model of Innovation.

Source: Rothwell and Zegveld [1994].

In short, what this model shows is that innovation is influenced by market forces and technological forces, which interact leading to the generation of new ideas. New needs reflect societal and market trends, such as new consumer tastes or demographic changes, while the state of production technologies and techniques establish new technological paradigms, which are additional sources of new ideas and applications.

However, it should be borne in mind that the dynamics of the model are still too sequential, since the generation of ideas leads to R&D, prototyping, production, and sales, giving feedback to the whole process with the generation of new ideas.

More satisfying models explicitly showed the steps in a chain of development from discovery to application and the marketplace, and when the diagram is supplemented by feedback arrows from one step to the previous one, then the flow of information indeed begins to look like links in a chain [Padmore *et al.*, 1998, p. 607].

In this direction, we highlight the contributions of Kline and Rosenberg [1986, p. 288], who offered the most recognized effort to move away from linear or sequential models, referring to this model as a chain-linked model. The authors criticize the linear model approach. "Even more important, from the viewpoint of understanding innovation, is the recognition that when the science is inadequate, or even totally lacking, we still can, do, and often have created important innovations and innumerable smaller, but cumulatively evolutionary changes." Kline and Rosenberg [1986] also saw that all of the relations in such a chain harness the existing corpus of knowledge or create new knowledge through research.

In this type of model, innovation is understood as a set of activities related to each other and whose results are often uncertain. Due to this uncertainty, there is no linear progression between the activities of the process and it is often necessary to return to previous phases to solve early-stage problems, so that at each stage it is possible to return to an earlier one. In this model, R&D is not a source of inventions, but a tool that is used to solve the problems that appear in any phase of the innovation process. The primary path of the innovation process is called the central chain of innovation, starting with design, development, and production to marketing activities (potential market, inventions, and/or product analytic design, detailed design and test, redesign, and production and, finally, distribution and marketing).

The company has a base of knowledge that is used to solve the problems that arise when innovating, while the research addresses the problems that cannot be solved with existing knowledge in order to expand that knowledge base and therefore to advance in the state of the art. According to the authors, R&D is not a precondition for innovation, but is added to it at any stage of the project. "We have already seen that the modern innovation is often possible without the accumulated knowledge of science and that explicit development work often points up the need for research, that is the new science.

Thus, the linkage from science to innovation is not solely or even preponderantly at the beginning of typical innovations" [Kline and Rosenberg, 1986, p. 390]. However, the paths include not only the central chain of innovation but also numerous feedback links that coordinate R&D with production and marketing, such as side-links to research, all along the central chain of innovation; long-range generic research to generate a backlog of innovations; potentiation of wholly new devices or processes from research, and the essential support of science itself from the products of innovation [Kline and Rosenberg, 1986, p. 303].

The OECD and Eurostat [2005, p. 24] cite the chain-link model as a useful approach to the innovation process. According to the *Oslo Manual*, the chain-link model "conceptualizes innovation in terms of interaction between market opportunities and the firm's knowledge base and capabilities. Each broad function involves a number of sub-processes, and their outcomes are highly uncertain. Accordingly, there is no simple progression; it is often necessary to go back to earlier stages in order to overcome difficulties in development. This means feedback between all parts of the process. A key element in determining the success (or failure) of an innovation project is the extent to which firms manage to maintain effective links between phases of the innovation process: the model emphasizes, for instance, the central importance of continuous interaction" (Fig. 2.4).

The chain-link model was a significant development showing a multidimensional approach to the innovation process. Thus, the model displays several feedback links among the stages in product development and sources of knowledge outside the firm. This approach provides a background for encouraging the points of the system to interact with commercial activities.

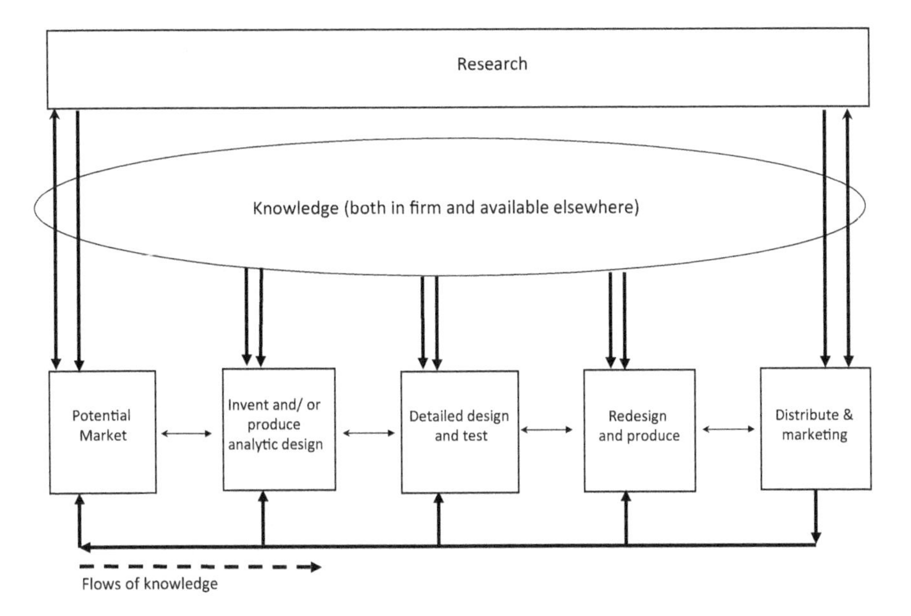

Figure 2.4. The Chain-Link Model of Innovation.

Source: Kline and Rosenberg [1986].

2.2.4 *The Fourth-Generation Innovation Process (Early 1980s to Early 1990s)*

Rothwell [1994] also defined a new generation of innovation models known as "integrated models." In this time period, the economy was characterized by a phase of recovery. The integrated models are based on studies that were carried out in the automotive and electronics industries in Japan in the mid-1980s, giving importance to just-in-time (JIT) methodologies deployed by these firms. At this time, there was a new generation of IT-based manufacturing and companies that tended to concentrate on their core business activities and critical technologies. Greater importance was placed on evolving generic technologies, with increased strategic emphasis on technological accumulation. Businesses also reached out to external parties for alliances that gave logic to the model. On the contrary, the shortening of the product life cycle made the time-to-market become a competitive element of the first magnitude.

Taking into account that time-to-market became an essential factor at a competitive level, the elements and phases of the process of technological innovation are managed, not from a linear point of view but as a series of processes that interact with each other, overlapping and simultaneous — all this while considering a greater integration of external actors, like the suppliers, within the development process, which is carried out in parallel. Two of the main features of leading Japanese companies are integration (suppliers and in-house departments in new product development) and parallel development, which became a design for manufacturing.

Although the mixed or interactive models incorporate retroactive communication processes between the various stages, they essentially remain sequential models, so that the beginning of a new stage is contingent upon the completion of the preceding stage. Apart from the consideration of the development time as a critical variable of the innovation process, the phases of the technological innovation process are beginning to be considered and managed, rather than through non-sequential processes, through overlapping or even concurrent or simultaneous processes.

The so-called "rugby approach" in product development contrasts with the traditional sequential approach and represents the idea of a group that, as a unit, tries to develop a distance, passing the ball backward and forward [Takeuchi and Nonaka, 1986]. Under this approach, the product development process takes place in a multidisciplinary group whose members work together from the beginning to the end. Instead of going through perfectly structured and defined stages, the process is shaped by interactions among the members of the group.

Rothwell [1994] referred to the innovation process as practiced in Nissan which is shown in Fig. 2.5 as a practical example of the integrated models. The so-called Schmidt–Tiedemann's model or concomitant model could be included among the integrated models. This is for certain authors one of the most practical models made to date [Forrest, 1991].

The model combines the three functional areas of the industrial innovation process: the research function (basic and applied), the technical function (technical evaluation, identification of know-how, and development needs), and the commercial function (market research,

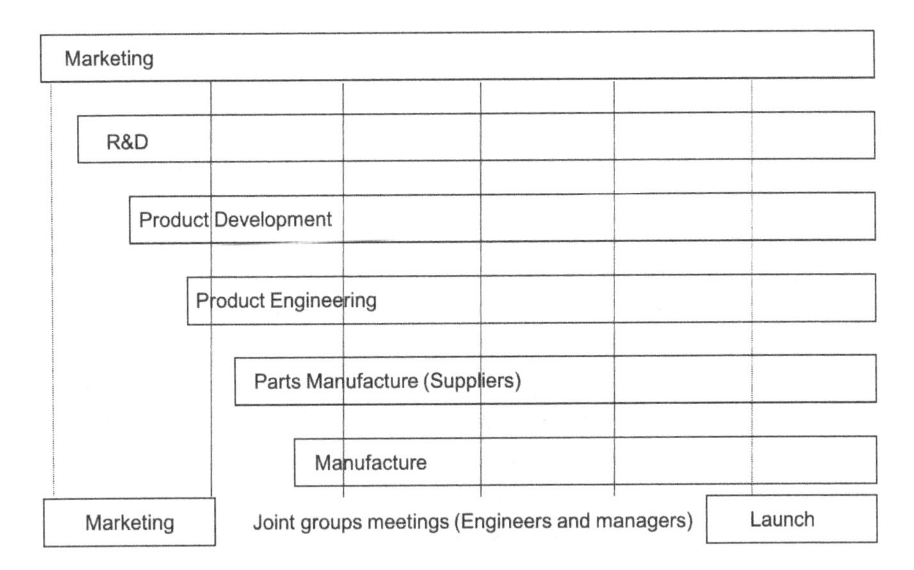

Figure 2.5. An Integrated (Fourth-Generation) Innovation Model.
Source: Graves [1987, cited in Rothwell, 1994, p. 12].

sales, and commercial). The concomitance model got its name because the research, commercial, and technical functions accompany each other throughout the innovation process with almost continuous interaction [Schmidt-Tiedemann, 1982].

2.2.5 *Toward the Fifth-Generation Innovation Process*

The world economy experienced rapid growth in the 1980s with high unemployment and business failure rates. This period was characterized by leading companies still looking for technology accumulation and a series of factors such as strategic networking, the importance of speed to market, better integrated products, and manufacturing, flexibility, and adaptability. All the aforementioned elements are sought as desired traits for business and product quality performance strategies. This model is the result of a series of global trends such as:

- increased strategic alliances and collaboration in R&D;
- integration of supply chains;

- relational networks of small and large companies and small businesses among themselves.

This kind of model is possible due to the use of electronic tools as support processes for innovation, which allow the implementation of network strategies. In this period, being a "fast innovator" is a cornerstone in the business strategy for the company's competitiveness, mainly in sectors with high technological change and short life cycles. Therefore, businesses seek to increase product development rates in order to be the first to market. Shorter time-to-market can have effects on market share, experience curve, monopoly profits, and customer satisfaction; however, the importance of being first and shortened development times can also have effects on development costs, which could be higher.

There is a U-shaped cost–time curve, which means that there is an optimum range of development where development costs are minimized. Rothwell affirmed that Japanese firms were operating close to the bottom of the curve, while US firms were less efficient. The author affirms that "Japanese firms are faster but more costly in their product development activities than their U.S. counterparts" [Rothwell, 1994, p. 15].

Leading innovators today adopt a variety of practices that are now shifting them toward a more favorable cost–time curve, with increased development speed and greater efficiency. These practices include internal organizational features, strong inter-firm vertical linkages, external horizontal linkages, and, more radically, the use of a sophisticated electronic toolkit, representing a shift in leading innovators toward the fifth-generation innovation process, which is defined by the author as a process of systems integration and networking (SIN).

Rothwell explained a number of factors that favor increasing development speed and efficiency such as explicit time-based strategy, top management commitment and support, adequate preparation, mobilizing commitment and resources, efficiency at indirect development activities, adopting a horizontal management style with increased decision-making at lower levels, committed and empowered product champions and project leaders, high-quality initial product specification (fewer unexpected changes), use of integrated (cross-functional) teams during development and prototyping (concurrent engineering), commitment to across-the-board quality control, incremental development strategy, adopting a "carry-over" strategy, product design combining the old with the new, designed-in flexibility, economy in technology, close linkages with primary suppliers, up-to-date component database, involving leading-edge users in design and development activities, accessing external know-how, use of computers for efficient intra-firm communication and data sharing, use of linked CAD systems along the production line (supplier, manufacturer, users), use of fast prototyping techniques, use of simulation modeling in place of prototypes, creating technology demonstrators as an input to simulation, and use of expert systems as a design aid [Rothwell, 1994, p. 15].

2.2.6 *Open Innovation Models*

The innovation landscape shifted dramatically with the beginning of the new millennium, requiring more collaboration and networking in the innovation realm. These new requirements respond to the globalization of the economy, an increasing competition, rapid changes in consumers' tastes and requirements, and the belief that there is relevant and valuable knowledge spread around the world. This new vision compels businesses to move from an innovation approach that looks inside the boundaries of the firm to another that is oriented beyond the firm's confines. This new vision has materialized in the open innovation models.

The concept of open innovation was coined by Chesbrough and is based on the idea that organizations increasingly rely on external innovation entities. Chesbrough [2006] refers to open innovation as "the use of purposive inflows and outflows of knowledge to accelerate internal innovation and to expand the markets for external use of innovation, respectively." Increasingly, companies are said to "openly" innovate by enlarging the process to include customers, suppliers, competitors, universities, research institutes, and others, as they rely on outside ideas for new products and processes [Granstrand, 2011].

Chesbrough recognizes the importance of networking and collaboration as the open and agile vehicles to put open innovation concepts into practice. Research and development is integrated beyond the boundaries of the firm, incorporating external knowledge into the innovation process and being more efficient. Through open innovation, the absorptive capacity of the firm and the development and diffusion of technology are enhanced. This model is based on the assumption that companies can tap into knowledge from outside the boundaries of the firm to deliver additional value to their customers, beyond that based solely on internal resources [Chesbrough, 2003, 2006].

Traditional models base innovation on the results of the research carried out internally by the R&D departments themselves, which in many cases, if these results do not fit with the strategies of the company itself, are abandoned or discarded, describing the innovation process as a funnel that tries to adjust the inputs to the company's own needs. However, within the so-called "open models," we face much more dynamic schemes that divulge from traditional linear conception and where the approaches from inside to outside or from outside to inside are those that dominate the logic of the system.

Chesbrough represented the model of open innovation according to the scheme we have included in Fig. 2.6.

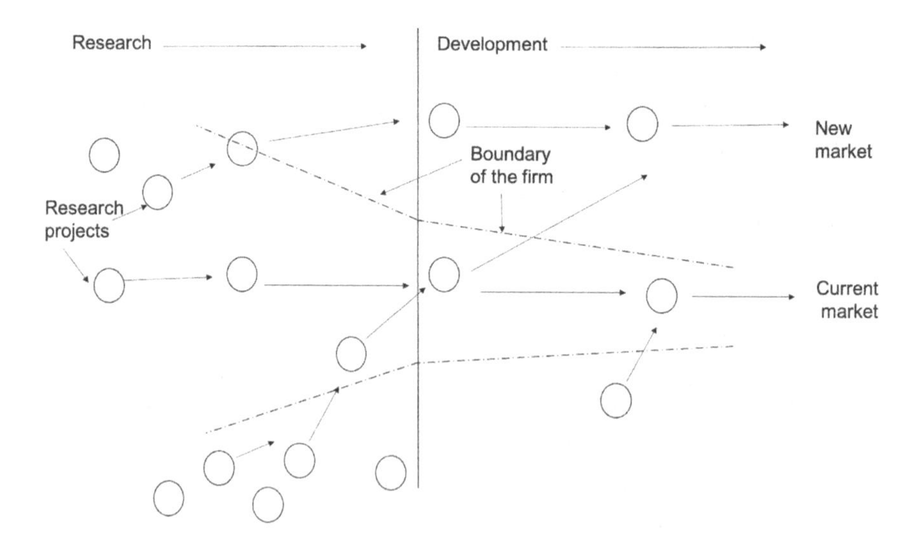

Figure 2.6. Open Innovation Models.

Source: Chesbrough [2006].

As symbolized in Fig 2.6, the innovation stream is represented as flowing through a funnel that is nourished by the creation of new knowledge through research projects. The funnel itself is divided into two steps, the research phase and the development phase, with the ultimate goal of reaching the market with new products or services or the implementation of new processes. Traditionally, there was no interaction between the inside and the outside of the funnel, managing the innovation process internally. However, with this new model, the boundaries of the firm have become permeable to the environment, allowing the relationship between the business and universities, research centers, other firms, consumers, and other related bodies. Thus, the firm can interact with its environment along the innovation process, importing new knowledge or exporting valuable knowledge produced by the firm.

REFERENCES

Abernathy, W. J., and Utterback, J. M. [1978]. Patterns of industrial innovation, *Technology Review*, 64(7), pp. 254–228.

Ahmed, P. K. [1998]. Culture and climate for innovation, *European Journal of Innovation Management*, 1(1), pp. 30–43.

Amar, A. D. [2004]. Motivating knowledge workers to innovate: A model integrating motivation dynamics and antecedents, *European Journal of Innovation Management*, 7(2), pp. 89–101.

Andriopoulos, C., and Lowe, A. [2000]. Enhancing organisational creativity: The process of perpetual challenging, *Management Decision*, 38(10), pp. 734–742.

Ayas, K. [1997]. *Design for Learning for Innovation* (Eburon, Delft).

Bennett, L. M., and Kerr, M. A. [1996]. A systems approach to the implementation of total quality management, *Total Quality Management*, 7(6), pp. 631–666.

Bessant, J., and Tidd, J. [2015]. *Innovation and Entrepreneurship*, 3rd edn. (Wiley, New Jersey).

Birkinshaw, J., and Hansen, M. T. [2007]. The innovation value chain, *Harvard Business Review*, 85(6), pp. 121–130.

Branscomb, L. M. [1993]. *Empowering Technology: Implementing a U.S. Strategy* (MIT Press, USA).

Brennan, A., and Dooley, L. [2005]. Networked creativity: A structured management framework for stimulating innovation, *Technovation*, 25(12), pp. 1388–1399.

Burgelman, R. A., Christensen, C. M., and Wheelwright, S. C. [2008]. *Strategic Management of Technology and Innovation*, 5th edn. (McGraw-Hill Education, Boston).

Burns, T., and Stalker, G. M. [1961]. *The Management of Innovation* (Tavistock Publications, London).

Chesbrough, H. W. [2003]. The era of open innovation, *MIT Sloan Management Review*, 44(3), pp. 35–41.

Chesbrough, H. W. [2006]. *Open Business Models: How to Thrive in the New Innovation Landscape*, 1st edn. (Harvard Business School Press, Boston).

Cohen, W. M., and Levinthal, D. A. [1990]. Absorptive capacity: A new perspective on learning and innovation, *Administrative Science Quarterly*, 35(1), pp. 128.

Cooper, R. G. [2011]. *Winning at New Products: Creating Value through Innovation*, 4th edn. (Basic Books, New York).

Cross, R. L., and Israelit, S. [2009]. *Strategic Learning in a Knowledge Economy* (Routledge, London).

Damanpour, F. [1987]. The adoption of technological, administrative, and ancillary innovations: Impact of organizational factors. *Journal of Management*, 13(4), pp. 675–688.

Davenport, T. H. [1995]. Will participative makeovers of business processes succeed where reengineering failed? *Planning Review*, 23(1), pp. 24–29.

Dewar, R. D., and Dutton, J. E. [1986]. The adoption of radical and incremental innovations: An empirical analysis, *Management Science*, 32(11), pp. 1422–1433.

Dosi, G. [1982]. Technological paradigms and technological trajectories: A suggested interpretation of the determinants and directions of technical change, *Research Policy*, 11(3), pp. 147–162.

Edquist, C. [1997]. *Systems of Innovation: Technologies, Institutions, and Organizations* (Psychology Press, London).

Forrest, J. F. [1991]. Practitioners' forum: Models of the process technological innovation, *Technology Analysis & Strategic Management*, 3(4), pp. 439–453.

Granstrand, O. [2011]. *The Economics of IP on the Context of Shifting Innovation Paradigm* (Chalmers Publication Library).

Graves, A. P. [1987]. Comparative trends in automotive research and development. International Motor Vehicle Programme, *Working Paper*, MIT, Cambridge, Massachusetts, USA.

Guimaraes, T., and Langley, K. [1994]. Developing innovation benchmarks: An Empirical Study, *Benchmarking for Quality Management & Technology*, 1(3), pp. 3–20.

Hayes. R. H., and Abernathy, W. J. [1980]. Managing our way to economic decline, *Harvard Business Review*, July–Aug, pp. 67–77.

Hidalgo, A., and Albors, J. [2008]. Innovation management techniques and tools: A review from theory and practice, *R&D Management*, 38(2), pp. 113–127.

Hobday, M. [2005]. Firm-level innovation models: Perspectives on research in developed and developing countries, *Technology Analysis & Strategic Management*, 17(2), pp. 121–146.

Hurley, R. F., and Hult, G. T. M. [1998]. Innovation, market orientation, and organizational learning: An integration and empirical examination, *The Journal of Marketing*, 62(3), pp. 42–54.

Hyland, P., and Beckett, R. [2005]. Engendering an innovative culture and maintaining operational balance, *Journal of Small Business and Enterprise Development*, 12(3), pp. 336–352.

Jantunen, A. [2005]. Knowledge-processing capabilities and innovative performance: An empirical study, *European Journal of Innovation Management*, 8(3), pp. 336–349.

Jones, G. R. [2000]. *Organizational Theory: Text and Cases*, 3rd edn. (Prentice Hall, Upper Saddle River).

Kahn, K. B. [2012]. *The PDMA Handbook of New Product Development*, 3rd edn. (Wiley, Hoboken)

Kandampully, J. [2002]. Innovation as the core competency of a service organisation: The role of technology, knowledge and networks, *European Journal of Innovation Management*, 5(1), pp. 18–26.

King, N., and Anderson, N. [1990]. Innovation in working groups, West, M. A., and Farr, J. L. (eds.), Chapter 4, *Innovation and Creativity At Work: Psychological And Organizational Strategies* (John Wiley & Sons, Oxford) pp. 81–100.

Kline, S., and Rosenberg, N. [1986]. An overview of innovation, Landau, R., and Rosenberg, N. (eds.), *The Positive Sum Strategy: Harnessing Technology for Economic Growth* (National Academy of Sciences, Washington) pp. 275–306.

Knight, K. E. [1967]. A descriptive model of the intra-firm innovation process, *The Journal of Business*, 40(4), p. 478.

Knight, R. M. [1987]. Corporate innovation and entrepreneurship: A Canadian study, *Journal of Product Innovation Management*, 4(4), pp. 284–297.

Koen, P. A., and Kohli, P. [1998]. Idea generation: Who has the most profitable ideas, *Engineering Management Journal*, 10(4), pp. 35–40.

Lendel, V., Hittmár, Š., and Siantová, E. [2015]. Management of innovation processes in company, *Procedia Economics and Finance*, 23, pp. 861–866.

Leonard-Barton, D. [1995]. *Wellsprings of Knowledge: Building and Sustaining the Sources of Innovation* (Harvard Business School Press, USA).

Loewe, P., and Dominiquini, J. [2006]. Overcoming the barriers to effective innovation, *Strategy & Leadership*, 34(1), pp. 24–31.

Mansfield, E. [1968]. *The Economics of Technological Change* (Norton, USA).

Martins, E. C., and Terblanche, F. [2003]. Building organizational culture that stimulates creativity and innovation, *European Journal of Innovation Management*, 6(1), pp. 64–74.

McAdam, R., and McClelland, J. [2002]. Individual and team-based idea generation within innovation management: Organizational and research agendas, *European Journal of Innovation Management*, 5(2), pp. 86–97.

Mintzberg, H. [1979]. *The Structuring of Organizations: A Synthesis of the Research* (Prentice-Hall, Englewood Cliffs, N.J.).

Mowery, D., and Rosenberg, N. [1979]. The influence of market demand upon innovation: A critical review of some recent empirical studies, *Research Policy*, 8(2), pp. 102–153.

Niosi, J. [1999]. Fourth-generation R&D: From linear models to flexible innovation, *Journal of Business Research*, 45(2), pp. 111–117.

Nonaka, I., and Takeuchi, H. [1995]. *The Knowledge-creating Company: How Japanese Companies Create the Dynamics of Innovation* (Oxford University Press).

OECD and Eurostat [2005]. *Oslo Manual* (OECD Publishing).

Padmore, T., Schuetze, H., Gibson, H., Padmore, T., Schuetze, H., and Gibson, H. [1998]. Modeling systems of innovation: An enterprise-centered view, *Research Policy*, 26(6), pp. 605–624.

Pavitt, K. [2002]. Innovating routines in the business firm: What corporate tasks should they be accomplishing? *Industrial and Corporate Change*, 11(1), pp. 117–133.

Petroni, A. [1998]. The analysis of dynamic capabilities in a competence-oriented organization, *Technovation*, 18(3), pp. 179–189.

Pisano, G. P. [1996]. *The Development Factory: Unlocking the Potential of Process Innovation* (Harvard Business Press)

Pohlmann, M., Gebhardt, C., and Etzkowitz, H. [2005]. The development of innovation systems and the art of innovation management — Strategy, control and the culture of innovation, *Technology Analysis & Strategic Management*, 17(1), pp. 1–7.

Rivas, R., and Gobeli, D. H. [2005]. Accelerating innovation at Hewlett-Packard, *Research-Technology Management*, 48(1), pp. 32–39.

Roffe, I. [1999]. Innovation and creativity in organizations: A review of the implications for training and development, *Journal of European Industrial Training*, 23(4/5), pp. 224–241.

Rothwell, R., Freeman, C., Horlsey, A., Jervis, V. T. P., Robertson, A. B., and Townsend, J. [1974]. SAPPHO updated — project SAPPHO phase II, *Research Policy*, 3(3), pp. 258–291.

Rothwell, R. [1994]. Towards the Fifth-generation Innovation Process, *International Marketing Review*, 11(1), pp. 7–31.

Rothwell, R., and Zegveld, W. [1985]. *Reindustrialization and Technology* (M.E. Sharpe).

Saad, M. [2000]. *Development through Technology Transfer: Creating New Organizational and Cultural Understanding* (Intellect, USA).

Saren, M. A. [1984]. A classification and review of models of the intra-firm innovation process, *R&D Management*, 14(1), pp. 11–24.

Schmidt-Tiedemann, K. J. [1982]. A new model of the innovation process, *Research Management*, 25(2), pp. 18–21.

Schmookler, J. [1962]. Changes in industry and in the state of knowledge as determinants of industrial invention, National Bureau of Economic Research, (ed.), *The Rate and Direction of Inventive Activity: Economic and Social Factors* (Princeton University Press, Princeton) pp. 195–232.

Senge, P. [1990]. *The Fifth Discipline* (Doubleday, New York).

Shipton, H., West, M. A., Dawson, J., Birdi, K., and Patterson, M. [2006]. HRM as a predictor of innovation: HRM as a predictor of innovation, *Human Resource Management Journal*, 16(1), pp. 3–27.

Smith, M., Busi, M., Ball, P., and Van Der Meer, R. [2008]. Factors influencing an organization ability to manage innovation: a structured literature review and conceptual model, *International Journal of Innovation Management*, 12(04), pp. 655–676.

Sørensen, J. B., and Stuart, T. E. [2000]. Aging, obsolescence, and organizational innovation, *Administrative Science Quarterly*, 45(1), pp. 81–112.

Takeuchi, H., and Nonaka, I. [1986]. The new product development game, *Harvard Business Review*, 64(1), pp. 137–146.

Tang, H. [1998]. An inventory of organizational innovativeness, *Technovation*, 19(1), pp. 41–51.

Tidd, J., and Bessant, J. [2013]. *Managing Innovation: Integrating Technological, Market and Organizational Change*, 5th edn. (John Wiley & Sons Inc., UK).

Tidd, J., and Hull, F. [2003]. *Service Innovation: Organizational Responses to Technological Opportunities & Market Imperatives* (Imperial College Press).

Trott, P. [2008]. *Innovation Management and New Product Development* (Pearson Education).

Usher, A. P. [2011]. *A History of Mechanical Inventions*, Revised edition (Dover Publications, New York).

Utterback, J. M. [1971]. The process of technological innovation within the firm, *Academy of Management Journal*, 14(1), pp. 75–88.

Utterback, J. M., and Abernathy, W. J. [1975]. A dynamic model of process and product innovation, *Omega*, 3(6), pp. 639–656.

Van de Ven, A. H. [1999]. *The Innovation Journey* (Oxford University Press).

Van de Ven, A. H., and Poole, M. S. [1990]. Methods for studying innovation development in the Minnesota innovation research program, *Organization Science*, 1(3), pp. 313–335.

van der Panne, G., van Beers, C., and Kleinknecht, A. [2003]. Success and failure of innovation: A literature review, *International Journal of Innovation Management*, 07(03), pp. 309–338.

Verloop, J. [2004]. *Insight in Innovation: Managing innovation by understanding the Laws of Innovation*, 1st edn. (Elsevier Science, Amsterdam).

Wheelwright, S. C. [2011]. *Revolutionizing Product Development: Quantum Leaps in Speed, Efficiency and Quality* (Free Press, New York).

White, M. A., and Bruton, G. D. [2010]. *The Management of Technology and Innovation: A Strategic Approach*, 2nd edn. (Cengage Learning, Inc., USA).

Woodman, R. W., Sawyer, J. E., and Griffin, R. W. [1993]. Toward a theory of organizational creativity, *Academy of Management Review*, 18(2), pp. 293–321.

Zwetsloot, G. [2001]. The management of innovation by frontrunner companies in environmental management and health and safety, *Environmental Management and Health*, 12(2), pp. 207–214.

CHAPTER 3

SOURCES OF INNOVATION: A STRATEGIC VIEW

3.1 WHERE DOES INNOVATION COME FROM?

Innovations arrive every day to the market in the form of new products or processes, and these innovations come from ideas that have been developed and implemented in the firm [Van de Ven, 1986]. If innovations rely on new ideas and the firm needs a sustainable stream of them [Boeddrich, 2004], it is important to analyze where ideas come from to know the sources on which innovations are based. The debate on strategies for sourcing innovation for the firm is not new, but nowadays it is gaining relevance due to the importance that these decisions have on the performance and competitiveness of the firm [Grimpe and Kaiser, 2010]. If innovativeness is linked to the development of new ideas, consequently, innovative firms are those that have the capacity to launch more and better ideas into the market than their competitors [Francis and Bessant, 2005].

Conventionally, the sources of innovation have been identified with the generation of new knowledge within the boundaries of the firm, usually through the R&D departments. This approach assumes a direct and positive relationship between in-house R&D expenditures and innovation performance, given the importance of knowledge generated in-house. Nevertheless, a broader approach has been adopted that includes not only traditional in-house sources of innovation but also those coming from the transfer of knowledge from other sources inside and outside the boundaries of the firm. Accordingly, knowledge transfer occurs between the traditional sources within the firm and external bodies; it can be purchased by the enterprise from external organizations to feed the innovation process, achieved through cooperation with outside R&D departments, or even produced from different departments within the firm. Therefore, innovation stems from either internal or external sources, the crucial point being that knowledge itself fuels innovation [Howells, 2002].

In earlier chapters, we discussed a two-fold orientation in the development of innovations starting either from the science or technology realm to reach the market, or stemming from the identification of a market need that has to be fulfilled by a new product or process. Both approaches have been shaped by the technology-push or market-pull models that we studied earlier in Chapter 1. While it seems to be a broadly accepted conclusion that customer needs are the origin of innovation, the technology approach also has many adherents. Correspondingly, Nye [2007, p. 2] argues, "Necessity is often not the mother of invention. In many cases, it surely has been just the opposite. When humans possess a tool, they excel at finding new uses for it. The tool often exists before the problem to be solved," propelling the importance of technology itself to shape new innovations.

The general approach to the study of the sources of innovation is also two-fold, considering internal and external sources to characterize the origin of innovations. On the one hand, there are sources that can be spotted inside the boundaries of the firm such as internal research and development (R&D) efforts, worker creativity, workforce training, new organizational methods, or the generation of new knowledge. On the other hand, it is also worth referring to external sources that exist beyond the firm's boundaries, such as suppliers, technology and machinery acquisition, collaboration with universities and research bodies, new clients' needs, changes in the market or the environment, extreme users who discover new applications and uses, competition, etc.

This categorization between external and internal technology sourcing has been investigated following the transaction costs approach and efficiency governance models [Mowery *et al.*, 1998; Robertson and Gatignon, 1998] and through the knowledge-based view theories [Grandori and Kogut, 2002; Kogut and Zander, 1992].

Moreover, we have to be aware that the sourcing strategy of the firm is not fixed, since the firm can vary the ways of accessing new innovation sources. It is not a question of deciding to adopt one approach exclusively, but to compound the best mix that serves the innovation strategy of the firm. Consequently, firms are moving from long-term to short-term decisions and to a more contingent approach to joint knowledge development about sourcing for innovation [Knudsen and Mortensen, 2011; Dittrich and Duysters, 2007]. All the aforementioned details rely on the fact that the pace of change in the marketplace is accelerating and businesses have to adapt their innovation strategy continuously. Innovation has broadened its sources from a more traditional R&D approach to a much larger array of potential sources, including employees, customers, collaborators, partners, and private inventors [Cooper and Edgett, 2008].

There are many authors who have studied the different types of innovation sources. For instance, Tidd and Bessant [2014, p. 98] pose a wide array of possible sources, among which we can cite shocks to the system (events that change the world and open up new ways of innovating), accidents (unexpected things that offer new innovation directions), watching others (imitation, copying, or improvements taking into account other's previous developments), recombination (adapting ideas and applications from one place and applying to another), regulation (changes in law that push innovation in new directions), covering latent needs in the market, inspiration, science-push (creating innovations stemming from new science developments), need-pull (innovations that are driven by new needs), users as innovators, and exploring new possibilities and future scenarios.

In any case, and coming back to the beginning of this chapter, we are going to use a simple structure to organize it, making a distinction between internal and external sources of innovation. We consider internal sources as all the in-house elements, resources, or events within the boundaries of the firm that could trigger innovation, while external sources of innovation are all the elements, resources, or events beyond the firm's boundaries.

3.2 INTERNAL SOURCES: RESOURCES AND CAPABILITIES OF THE FIRM

3.2.1 *Knowledge and Organizational Learning*

There are many authors who describe innovation as an organizational learning process [Cohen and Levinthal, 1990; Dodgson, 1991; Hitt *et al.*, 2000].

From this perspective, the firm learns to improve its competences, which is the basis for its competitive advantage, taking into account that the capabilities of the firm must change and adapt to the new requirements of the market. The need for new knowledge to fulfil new market issues has been triggered by the increasing competition and the pace of the technological change. Knowledge has become the most important competitive resource nowadays, being the most valuable asset of the company.

According to Henderson and Cockburn [1996], the theory of organizational learning defends that organizational innovation performance relies on the knowledge base of the firm. Considering knowledge as the main contingency variable, many authors have examined the impact on the firm of choosing between internal or external knowledge [Kogut and Zander, 1992; Grandori and Kogut, 2002], which is especially relevant in technological development [Birkinshaw *et al.*, 2002]. Taking into account this approach, innovation is considered a dependent outcome of an ongoing transfer and assimilation of knowledge [Cantner *et al.*, 2011]. Therefore, the generation of new knowledge or the transfer of existing knowledge from other sources fuels innovation in the firm.

Firms often rely on their R&D departments to create new knowledge that could serve as a primary source of innovation [Hirsch-Kreinsen *et al.*, 2005]. This organizational learning in the firm fuels the knowledge that the business needs to install, operate, maintain, improve, and develop its technology [Hamel and Prahalad, 1994]. The company's own R&D efforts are developed in order to produce distinctive and innovative knowledge, which can create absorptive capabilities [Cohen and Levinthal, 1990; Jansen *et al.*, 2006]. Thus, it is widely assumed that the innovativeness and competitiveness of the firm depend on the R&D intensity of the business; it is the investment in research and development. Frenz and Ietto-Gillies [2009], analyzing the Community Innovation Survey for the UK, argue that internal sources of innovation such as intra-company knowledge sources, internal generation, and brought-in R&D make a difference in terms of innovation performance with external and cooperation-based innovation efforts. It is also evident that the importance of the existence of internal R&D sources depends on the kind of industry. For instance, science-based and high-technological industries such as electronics, aircraft and spacecraft, communication equipment, chemical, machinery and equipment, or pharmaceuticals are heavily dependent on their ability to come out with new discoveries and inventions to fuel innovation processes. In this kind of industry, it is more common to find internal R&D facilities leading the generation of new knowledge.

Nevertheless, according to Dosi [1988], innovation is not only about R&D, since it includes discovery, development, imitation, and adoption of new products or processes. Accordingly, Rosenberg [1982] argues that innovation stems from other activities beyond those of R&D and that many firms, for instance, those from low- and medium-tech sectors, such as building, rubber, and plastics or basic metals, do not work on the latest scientific or technological knowledge, but, rather, transform existing knowledge into economically viable knowledge [Bender and Laestadius, 2005], which can include learning, appraisal, and evaluation of technologies and internal experimentation. This approach also depends on the innovation strategy of the firm, since innovation leaders normally make a difference in R&D, with high investment in research and development, while businesses that act as innovation followers are keener to leave R&D leadership to other firms while they focus their efforts on design, development, and implementation, many times through reverse engineering.

Another point to consider is the role of research and development in the services sector. There is evidence that R&D is less important in service sector firms than in industrial business, no matter the type and subsector, with the exception of knowledge-intensive business services (KIBS) that include activities such as engineering, research and development, telecommunication, and software, which are R&D intensive [García Manjón, 2008]. However, technology also plays an important role in service firms and must not be neglected. This is the case of information and communication technologies (ICT), which are baseline technologies for a wide array of service sectors and support and an endless number of innovations in the services arena.

3.2.2 *Innovation by Design*

Another important topic to be considered as a source for innovation is design. Design is defined as a creative process that can have a rational, innovative, or artistic orientation [Bender and Laestadius, 2005] that also covers a wide variety of activities such as architecture, fashion, and interior, graphic, industrial, or engineering design [Walsh, 1996]. The term design also refers to the stages needed to move from the prototyping phase into manufacturing [Marsili and Salter, 2006]. There are many authors in the literature who have moved from the technology-driven approach to innovation to underlining the importance of meaning-driving products, services, and experiences [Verganti, 2008; Geels, 2004], where the concept of design is intimately related to the meaning that people give to products [Battistella *et al.*, 2012].

According to different investigations, design is an important source of knowledge and learning in the firm, involving its own organizational structure and processes and other firm's functions as well as interaction with external bodies [Verona and Ravasi, 2003]. Design has also been analyzed as a contributor to innovation outputs [Marsili and Salter, 2006; Walsh, 1996; Verganti, 2008] and it seems to have special importance in low- and medium-tech industries, where design and quality are the most important ways of competing [Hansen and Serin, 1997]. Design is also considered a central competence to leverage performance and promote product and business innovation [Bertola and Teixeira, 2003]. This design-driven innovation view stems from approaches in which "form" has meaning, since people perceive not only forms but also the meaning associated with them — making sense of things [Krippendorff, 1989].

According to Filippetti [2011] who studied design as a source of innovation, the author raises the finding that design seems to be a complementary source of innovation more than an alternative one, pointing to the concept that design as a source of innovation is more important in those firms characterized by a complex strategy and with an attitude of openness to the environment in terms of innovation, and being associated with better economic performance. Design also plays an important role in shaping customers' needs through the functionalities and uses of the product and going beyond the technology that supports it.

3.2.3 *Employees as Innovators*

Employees are also another significant source for innovation, from the time the internal ideas trigger the innovation process itself; considering creativity is the main individual outcome that plays a role in the organizational context. In literature there are researchers who give a central role to employees, identifying individuals as the ultimate source of creativity and innovation [Amabile *et al.*, 1996]. At the individual level, innovation starts with the generation of new ideas from an employee [Kanter, 1988a]. Employee-driven innovation means that employees contribute actively and systematically to the innovation process.

The generation of creative ideas by employees is often related to pitfalls or problems at work that challenge employees [Kanter, 1988a; Drucker, 1985]. Some authors defined employee-driven innovation as a new idea created by employees that results in a new, shared, and sustainable routine to overcome problems experienced inside the firm [Feldman and Pentland, 2003]. This approach leans on the research that argues that employees apply innovation as a problem-solving-based strategy in order to cope with work-related problems and biases that provoke strains in the workplace [Janssen, 2000]. This kind of idea varies from incremental or evolutionary changes in the firm to radical ideas that lead organizations to more profound changes and a paradigm shift [Kanter, 1988b]. In the same vein, ideas can vary from those that are focused on the core activity of the firm and those who operate outside the main activities of the business. Depending on that, innovation will be more or less pervasive, since ideas aiming at the core of the business derive from changes affecting the business logic, and those focusing on non-central activities entail little improvements or changes in the business. Thus, employee-driven innovation has a lot to do with organizational learning and the innovation is based on practices in the workplace. This practice-based innovation as it is referred to by Melkas and Harmaakorpi [2011, p. 163] means that the "learning and innovation process begins with questioning, a disturbance or the emergence of a problematic situation in the conduct of a task or in the interplay with other people."

The literature also enhances the importance of employee-based innovation, since it may be more productive than user-based innovations; those ideas from employees are more feasible than ideas from other groups [Poetz and Schreier, 2012] because it is easier to transform employee ideas into commercial and viable products, as employees have a deep knowledge of the technology and products are more viable for the firm

[Magnusson, 2009]. It is also worth citing the special interests of those employees who are at the same time users of the product as a source of innovation because this group represents a confluence of valuable knowledge, user experience, and employee perspective [Harrison and Corley, 2011]. These kinds of employees can assume different roles in the innovation process, such as idea generation, product testing, or marketing experts [Wadell *et al.*, 2013].

Employee participation in many of the basic management principles that are focused on seeking continuous improvement in the organization plays an important role in promoting innovation [Fuller *et al.*, 2006]; for instance, in Kaizen-based programs [Imai, 1986] or suggestion and participation programs for employees [Unsworth, 2001].

Taking for granted that employee participation is valuable for promoting innovative output in the organization, there is an increasing interest in studying the factors that trigger innovative behavior of employees that are significant in identifying the antecedents of this innovative work behavior [Jansen *et al.*, 2006]. Some of the most important features are employee traits, values, or abilities. Thus, some authors argue that among these traits the entrepreneurial profile of the employee and the capacity to assume risks associated with innovation are facilitators of innovation [Kanter, 1985]. Moreover, Cerinsek and Dolinsek [2009] posit some underlying characteristics of the individual such as curiosity, autonomy, flexibility, perceptiveness, motivation, ambitiousness, creativity, self-confidence, and entrepreneurship as moderators of innovative behavior. In this vein, Janssen [2003] argues that innovative workers are even more likely to take part in conflict if changes in the work environment are central to their sense of identity, perceiving the resistance of co-workers as a relevant source of divergence.

Another crucial point for enabling employee participation in the innovation process is the role of supervisors, since the generation and implementation of new ideas by employees depend on the supportive style by the supervisor [Axtell *et al.*, 2000] and the formal structure and organizational rationale in the firm. Thus, mechanistic, rigid organizational structures are more likely to generate conflicts in periods of change than flexible and open structures that more quickly adapt innovatively to changing situations [Mintzberg, 1979]. This seems to be logical, as innovation is a risky activity and the support and guidance of supervisors are essential to support employee initiatives.

There are many examples of the involvement of employees to generate ideas and spark innovation. Thus, following the examples set by Morgan [2014], we can cite Toyota's seminal initiative with the "Creative Ideas Suggestive System" in 1951, which has contributed more than 40 million ideas so far. Another good example is Whirlpool, the home appliance manufacturer with over 70,000 employees around the world. They have created a program called "Idea Labs" that allows employees to come forward with an idea via online platforms. At the end of the year 2014, they got a shift in sales from a declining figure of 2% to a 2% increase instead. The Ideaboxes program from Ericsson is also well known, as it matches employee's ideas with managers who seek particular needs in the company. The program is supported by an internal team called "Innova" acting as an internal venture team for funding start-up capital to employees with good ideas.

3.2.4 *Training to Fuel the Learning Process in the Firm*

Continuing with internal sources of innovation, we can cite training as an important activity that is linked to innovation performance [Dodgson and Rothwell, 1995]. As we consider knowledge the most important asset in the firm for increasing strategic capabilities, training is a way to raise the knowledge base of the firm, increasing the value of human capital and its absorptive capacity [Cohen and Levinthal, 1990]. Learning promotes the creation, exchange, and use of information and knowledge that impact innovation performance [Hatch and Dyer, 2004].

Training programs can impact different stages of the innovation process from idea generation through promotion and implementation [Nonaka and Takeuchi, 1995]. Accordingly, if the identification and effective transfer and absorption of knowledge are required for innovation [Chen and Huang, 2009], then training is expected to increase the innovative performance of the firm; however, this only occurs when training prompts an increase in sharing and exploitation of knowledge among employees [Kang *et al.*, 2007].

3.3 EXTERNAL SOURCES: GENERAL ENVIRONMENT AND MARKET

3.3.1 *External Sources of Knowledge*

As mentioned earlier, the generation of new knowledge is a key factor in competing in a world driven by a hectic technological pace and increasing competition. Taking into account the importance of the knowledge generated in-house through the R&D departments and other means, it is also necessary to consider the interaction of the firm within its own environment to identify sources of knowledge, often in the area in which the firm operates. Given the increasing uncertainty, firms explore new forms of innovation, becoming increasingly interested in cooperating with other firms in order to reach, develop, and exchange the knowledge necessary for the activity of the firm. This is a way to lower the firm's risk in developing its innovation strategy, considering it is no longer an individual or isolated activity but a collaborative one. More and more, ideas are developed in an exchange or collaborative process with an established set of partners and networks, where the innovative power of each point of the network becomes decisive for the others. Collaboration for common R&D and innovation efforts is not new in the innovation arena. Since the 1980s, there has been evidence of inter-firm R&D alliances [Hagedoorn, 2002] and collaboration in knowledge exchange and innovation [Tether, 2002; Laursen and Salter, 2004]. Due to increasing competition, closed internal R&D teams are shifting rapidly toward more open and collaborative R&D and innovation processes, which are harnessing the potential of the knowledge outside the firm to build alliances, technological agreements, and licensing schemes [Hagedoorn, 2002].

Accordingly, some authors consider that the success of technological innovations relies on the ability of a firm to identify, utilize, and exploit external knowledge [Lin *et al.*, 2002] and the acquisition of external knowledge bases [Cohen and Levinthal, 1989; Huber, 1991]. This ability depends on technological competences that are applied to tap into the external sources of knowledge [Vega-Jurado *et al.*, 2008]. Focusing on the role of R&D and its role on innovation, some authors point out that, to a large extent, the knowledge acquired comes from outside formal R&D systems [Baldwin and Hanel, 2003]. Firms can acquire knowledge via cooperative agreements and with the participation of public and private bodies, including firms, universities, and public institutions [Frenz and Ietto-Gillies, 2009].

The process of knowledge transfer depends on multiple factors that ease the process of integration of external knowledge into the firm. First and foremost is the nature of the knowledge itself; the management of knowledge transfer differs depending on whether it is tacit or explicit knowledge [Dhanaraj *et al.*, 2004]. Tacit knowledge is normally transferred by non-standardized and tailored processes [Polanyi, 1966], whereas explicit knowledge is embedded in standardized procedures and products [Martin and Salomon, 2003]. It is not odd to find that integrating talented people from other companies is the way to incorporate the tacit knowledge that they treasure. Second, we refer to the characteristics of the recipient firm, depending on its absorptive capacity, which constitutes the "ability to recognize the value of new information, assimilate it, and apply it to commercial ends" [Cohen and Levinthal, 1990, p. 128]. Finally, we consider the embeddedness of the firm either in the environment of the firm or the corporate networks it belongs to [Andersson *et al.*, 2005]. Thus, the deeper the integration of the firm in the environment or in corporate networks, the easier the transferability process is.

Many times, there are barriers that prevent firms from reaching the sources of knowledge directly, maybe due to its size or simply due to the lack of ability. Gaps between suppliers and users of technology demand intermediaries to facilitate the absorption of knowledge, which is especially relevant in the case of small- and medium-sized businesses. Accordingly, researchers have also pointed out the importance of the existence of knowledge providers as an intermediary between knowledge sources and beneficiaries. These kinds of intermediaries are defined as knowledge-intensive business services (KIBS), which are composed of consultancy firms, research institutes, universities, and engineering firms. Frenz and Ietto-Gillies [2009] note that KIBS' enhance the international dimension of internal networks and the collaboration processes between internal and external sources to generate new knowledge, which leverages the innovation potential of the firm.

Cooperation has been analyzed under the transaction cost approach considering the sharing of costs and risk of R&D activities and the dissemination of results [Williamson, 1985]. The strategic approach considers R&D collaboration as a way to incorporate unique knowledge to build up competitive advantages [Teece, 1996], or spillovers among partners, considering the spillover as an incentive to cooperate [Petit and Tolwinski, 1999]. Cooperation can be undertaken with suppliers, customers, competitors, firms that belong to the same groups, universities, and public centers of R&D. Nevertheless, following the transaction cost theories, there are some barriers to external knowledge sourcing, such as the selection and coordination with external partners [Steensma and Corley, 2001] and the risks of imitation and knowledge protection [Becerra *et al.*, 2008]. These barriers may be offset by cost-sharing effects and lowering risk levels in knowledge absorption [Huang *et al.*, 2009].

Knowledge transfer from bodies outside the boundaries of the firm is normally a two-way interaction process by means of exchange and collaboration. This exchange process comes either from contractual agreements with knowledge providers or from knowledge "spillovers," which is the appropriation of knowledge that is available in the environment as a result of innovation efforts from other firms or innovation agents. Spillovers as a source of innovation is not so much the result of a voluntary exchange, but the consequence of unintentional leakages of knowledge [De Bondt, 1997]. Spillovers can come from competitors, suppliers, users, or other specialized knowledge producers such as universities. All this can include the acquisition of knowledge by means of patent purchasing or license agreements, external R&D done in collaboration with other firms or institutions, and R&D carried out by external bodies on behalf of the enterprise.

3.3.2 *Suppliers as a Source of Innovation*

The supply chain is another relevant player that is involved in the innovation process of the firm, which sometimes is crucial for its development [Schiele, 2006]. Pavitt already studied the supplier-dominated industries and posits that "in sectors made up of supplier dominated firms, we would expect a relatively high proportion of the process innovations used in the sectors to be produced by other sectors, even though a relatively high proportion of innovative activities are directed to process innovations" [Pavitt, 1984, p. 356]. Therefore, the author relates to the interaction between sectors to promote innovativeness. Some industries, mainly those with low technological intensity, rely on suppliers as the first source to promote innovation, granting their suppliers the role of technological partners in the innovation strategy of the firm.

Some authors in argue that cooperation with suppliers is crucial for innovation performance. In some cases, they consider collaboration with suppliers to be more determinant than technological competence to guarantee the success of the developments of products [McCutcheon *et al.*, 1997]. Other analysis also reports a clear connection between customer–supplier relationships with innovation performance [Cantista and Tylecote, 2008]. Accordingly, it seems to be a link between the purchasing function of the firm and its innovation role. This is a bi-directional process; on the one hand, the customer may have an idea that is submitted to the supplier for its consideration and advise on how to implement it, and on the other hand, the supplier itself may contribute new ideas to the customer [Hakansson and Eriksson, 1993] generating a coupling effect that is beneficial for both.

The innovation process based on relationships with suppliers also takes different approaches. The first one is characterized by short-term relationships primarily based on price and with a low level of involvement by both parties, which has been referred to as the competitive model in the literature [Hendrick and Ellram, 1993]. This approach is far from an authentic strategic partnership between suppliers and the firm in the field of innovation. The second approach is a long-term relationship, based on a continuous exchange of information and high involvement on both operational and strategic aspects. Some authors have referred to this model as customer–supplier partnership [Katz and Shapiro, 1985], which is closer to the strategic point of view of the firm. From the innovation point of view, the relationship established between suppliers and client firms is crucial in determining the effects on innovation performance. Thus, aspects such as the characteristics of the cooperation between suppliers and firms, the exchange of information, the generation of trust in the supplier's abilities to respond to the needs of the firm, and the level of investment in specific assets are key points that determine the effects on the innovation process [Carr and Kaynak, 2007; Groznik and Maslaric, 2010; Sivadas and Dwyer, 2000; Chiara Di Guardo and Valentini, 2007].

3.3.3 *Innovations Coming from Users*

Another external source of innovation may come from users [Jeppesen and Frederiksen, 2006] who are an important and prolific source of innovation and ideas for the firm [Bogers *et al.*, 2010; Poetz and Schreier, 2012]. Users' ideas may be valuable due to their novelty, added value, and market potential [Magnusson, 2009], in many cases greater than those from employees [Poetz and Schreier, 2012], and the motivation of users to adopt it [von Hippel, 2005]. The literature has widely studied the role of users in the generation of innovations to improve products and services [von Hippel, 1976; Herstatt and von Hippel, 1992; Jeppesen and Frederiksen, 2006], which has gained importance though the years and in a wide range of industries, such as the oil industry [Enos, 1962], chemical industry [Freeman *et al.*, 1968], scientific instruments where users were behind 80% of products innovations [Urban and von Hippel, 1988], consumer goods [Franke and Shah, 2003], or industrial products [Morrison *et al.*, 2004].

Nevertheless, firms struggle to incorporate this value from users because they are outside the boundaries of the firm, and users and corporate business are two separate worlds. Among these barriers we recognize the difficulty in identifying innovative users and their inability to formulate ideas properly, the interpretation of user ideas [Mahr and Lievens, 2012] that can sometimes generate biases, the lack of user knowledge specifically about organizational processes [Homburg *et al.*, 2009], the lack of incentives for sharing [de Jong *et al.*, 2015], and the proliferation of ideas that do not fit the organization [Poetz and Schreier, 2012].

Furthermore, awareness of users' ideas doesn't guarantee the incorporation of valuable ideas within the firm. Special innovative ideas occur among lead users of the firm's product and services [von Hippel, 1986] from the time when these kind of users are the vanguard of the market and experience needs that other users do not, and they gain more value when fulfilling these needs [Schweisfurth, 2017], being more likely to develop more market-oriented innovations than others [Franke *et al.*, 2006].

Following von Hippel [2009, p. 7], "Since lead users are the leading edge of the markets with respect to important market trends, one can guess that many of the novel products they develop for their own use will appeal to other users too and so might provide the basis for product [that] manufacturers would wish to commercialize." Accordingly, von Hippel credits users with the capacity to envision new products or functions that may be useful in the future for the vast majority of the market.

The involvement of lead users in the innovation process appears to be correlated with the success of the innovation in the market. Consequently, Franke and von Hippel [2003], analyzing the software industry, determined that as the innovator displays more lead-user characteristics, so is the attractiveness of the innovation developed. It seems to be logical that when businesses and users go hand-in-hand in development of innovations, greater market acceptance ensues. von Hippel [2009] argues that lead users display specific needs that are different from mass market participants. Thus, adds the author, mass manufacturers design their products to fulfil general market needs for the majority of users, which does not sufficiently satisfy a specific group's needs such as those of lead users.

Another interesting point is the spillover effect that is produced when users are involved in the development of innovation. Accordingly, users often reveal and disseminate their innovations freely to the public. While producers partially reveal the innovations contained in a new product or service, empirical research shows that users freely reveal their innovations, giving away all the intellectual property rights voluntarily, and all the interested parties have access to this information [Harhoff *et al.*, 2003]. The authors propose that free revealing is a common phenomenon that could be profitable considering the different agents that operate in the market. According to von Hippel [2009], free revealing seems to be a valuable source of innovation. As the author argues, free revealing has occurred throughout history and not only in well-known examples such as the "open-source" software development community where users systematically reveal and freely share the code at their personal expense [Raymond, 1999]; but we can refer also to the iron industry and the collective invention [Allen, 1983], pumping engine industry [Nuvolari, 2004], and sporting equipment [Franke and von Hippel, 2003] among others, all cited by Hippel [2009].

Another relevant point in the study of innovation sources is the geographical variable. Asheim and Gertler [2005] and Cantwell and Iammarino [2000] argue that the geographic context of the firm constitutes a relevant factor for both the generation and transfer of knowledge. The existence of diverse geographical linkages of the firm, whether they are internal or external to the company itself, ease the company's reach into different knowledge environments and the transferability process [Frenz and Ietto-Gillies, 2009].

The spillover effects tend to increase when the company has geographical diversity, leveraging knowledge transfer and innovation opportunities in the various locations where the firm operates [Cantwell, 1989], which impacts technological learning and performance [Zahra *et al.*, 2000].

REFERENCES

Allen, R. C. [1983]. Collective invention, *Journal of Economic Behavior & Organization*, 4(1), pp. 1–24.

Amabile, T. M., Conti, R., Coon, H., Lazenby, J., and Herron, M. [1996]. Assessing the work environment for creativity, *Academy of Management Journal*, 39(5), pp. 1154–1184.

Andersson, U., Björkman, I., and Forsgren, M. [2005]. Managing subsidiary knowledge creation: The effect of control mechanisms on subsidiary local embeddedness, *International Business Review*, 14(5), pp. 521–538.

Asheim, B. T., and Gertler, M. S. [2005]. *The Oxford handbook of innovation,* Fagerberg, J. and Mower, D. C. (eds.), Chapter 11, *The Geography Of Innovation: Regional Innovation Systems* (Oxford Publishing, Oxford) pp. 291–317.

Axtell, C. M., Holman, D. J., Unsworth, K. L., Wall, T. D., Waterson, P. E., and Harrington, E. [2000]. Shopfloor innovation: Facilitating the suggestion and implementation of ideas, *Journal of Occupational and Organizational Psychology*, 73(3), pp. 265–285.

Baldwin, J. R., and Hanel, P. [2003]. *Innovation and Knowledge Creation in an Open Economy: Canadian Industry and International Implications* (Cambridge University Press, USA).

Battistella, C., Biotto, G., and De Toni, A. F. [2012]. From design driven innovation to meaning strategy, *Management Decision*, 50(4), pp. 718–743.

Becerra, M., Lunnan, R., and Huemer, L. [2008]. Trustworthiness, risk, and the transfer of tacit and explicit knowledge between alliance partners, *Journal of Management Studies*, 45(4), pp. 691–713.

Bender, G., and Laestadius, S. [2005]. Non-science based innovativeness. On capabilities, relevant to generate profitable novelty, *Journal of Mental Changes*, 11(1–2), pp. 123–170.

Bertola, P., and Teixeira, J. C. [2003]. Design as a knowledge agent: How design as a knowledge process is embedded into organizations to foster innovation, *Design Studies*, 24(2), pp. 181–194.

Birkinshaw, J., Nobel, R., and Ridderstråle, J. [2002]. Knowledge as a contingency variable: Do the characteristics of knowledge predict organization structure? *Organization Science*, 13(3), pp. 274–289.

Boeddrich, H.-J. [2004]. Ideas in the workplace: A new approach towards organizing the fuzzy front end of the innovation process, *Creativity and Innovation Management*, 13(4), pp. 274–285.

Bogers, M., Afuah, A., and Bastian, B. [2010]. Users as innovators: A review, critique, and future research directions, *Journal of Management*, 36(4), pp. 857–875.

Cantista, I., and Tylecote, A. [2008]. Industrial innovation, corporate governance and supplier–customer relationships, *Journal of Manufacturing Technology Management*, 19(5), pp. 576–590.

Cantner, U., Joel, K., and Schmidt, T. [2011]. The effects of knowledge management on innovative success — An empirical analysis of German firms, *Research Policy*, 40(10), pp. 1453–1462.

Cantwell, J. [1989]. *Technological Innovation and Multinational Corporations* (Blackwell, Oxford).

Cantwell, J., and Iammarino, S. [2000]. Multinational corporations and the location of technological innovation in the UK regions, *Regional Studies*, 34(4), pp. 317–332.

Carr, A. S., and Kaynak, H. [2007]. Communication methods, information sharing, supplier development and performance: An empirical study of their relationships, *International Journal of Operations & Production Management*, 27(4), pp. 346–370.

Cerinsek, G., and Dolinsek, S. [2009]. Identifying employees' innovation competency in organisations, *International Journal of Innovation and Learning*, 6(2), pp. 164–177.

Chen, C. J., and Huang, J. W. [2009]. Strategic human resource practices and innovation performance — The mediating role of knowledge management capacity, *Journal of Business Research*, 62(1), pp. 104–114.

Chiara Di Guardo, M., and Valentini, G. [2007]. Explaining the effect of M&A on technological performance, Cooper, C. and Finkelstein, S. (eds.), *Advances in Mergers and Acquisitions* (Emerald Group Publishing Limited, UK) pp. 107–125.

Cohen, W. M., and Levinthal, D. A. [1989]. Innovation and learning: The two faces of R & D, *The Economic Journal*, 99(397), pp. 569–596.

Cohen, W. M., and Levinthal, D. A. [1990]. Absorptive capacity: A new perspective on learning and innovation, *Administrative Science Quarterly*, 35(1), p. 128.

Cooper, R. G., and Edgett, S. [2008]. Ideation for product innovation: What are the best methods, *PDMA Visions Magazine*, 1(1), pp. 12–17.

De Bondt, R. [1997]. Spillovers and innovative activities, *International Journal of Industrial Organization*, 15(1), pp. 1–28.

de Jong, J. P. J., von Hippel, E., Gault, F., Kuusisto, J., and Raasch, C. [2015]. Market failure in the diffusion of consumer-developed innovations: Patterns in Finland, *Research Policy*, 44(10), pp. 1856–1865.

Dhanaraj, C., Lyles, M. A., Steensma, H. K., and Tihanyi, L. [2004]. Managing tacit and explicit knowledge transfer in IJVs: The role of relational embeddedness and the impact on performance, *Journal of International Business Studies*, 35(5), pp. 428–442.

Dittrich, K., and Duysters, G. [2007]. Networking as a means to strategy change: The case of open innovation in mobile telephony, *Journal of Product Innovation Management*, 24(6), pp. 510–521.

Dodgson, M. [1991]. Technology learning, technology strategy and competitive pressures, *British Journal of Management*, 2(3), pp. 133–149.

Dodgson, M., and Rothwell, R. [1995]. *The Handbook of Industrial Innovation* (Edward Elgar Publishing, Cheltenham, UK).

Dosi, G. [1988]. Sources, procedures, and microeconomic effects of innovation, *Journal of Economic Literature*, pp. 1120–1171.

Drucker, P. F. [1985]. The discipline of innovation, *Harvard Business Review*, 63(3), pp. 67–72.

Enos, J. L. [1962]. Invention and innovation in the petroleum refining industry, National Bureau of Economic Research (ed.), *The Rate and Direction of Inventive Activity: Economic and Social Factors* (Princeton University Press, Princeton) pp. 299–322.

Feldman, M. S., and Pentland, B. T. [2003]. Reconceptualizing organizational routines as a source of flexibility and change, *Administrative Science Quarterly*, 48(1), pp. 94–118.

Filippetti, A. [2011]. Innovation modes and design as a source of innovation: A firm-level analysis, *European Journal of Innovation Management*, 14(1), pp. 5–26.

Francis, D., and Bessant, J. [2005]. Targeting innovation and implications for capability development, *Technovation*, 25(3), pp. 171–183.

Franke, N., and Shah, S. [2003]. How communities support innovative activities: An exploration of assistance and sharing among end-users, *Research Policy*, 32(1), pp. 157–178.

Franke, N., and von Hippel, E. [2003]. Satisfying heterogeneous user needs via innovation toolkits: The case of Apache security software, *Research Policy*, 32(7), pp. 1199–1215.

Franke, N., von Hippel, E., and Schreier, M. [2006]. Finding commercially attractive user innovations: A test of lead-user theory, *Journal of Product Innovation Management*, 23(4), pp. 301–315.

Freeman, C., Robertson, A. B., Whittaker, P. J., Curnow, R. C., Fuller, J. K., Hanna, S. [1968]. Chemical process plant: Innovation and the world market, *National Institute Economic Review*. pp. 29–57.

Frenz, M., and Ietto-Gillies, G. [2009]. The impact on innovation performance of different sources of knowledge: Evidence from the UK community innovation survey, *Research Policy*, 38(7), pp. 1125–1135.

Fuller, J. B., Marler, L. E., and Hester, K. [2006]. Promoting felt responsibility for constructive change and proactive behavior: Exploring aspects of an elaborated model of work design, *Journal of Organizational Behavior*, 27(8), pp. 1089–1120.

García Manj6n, J. V. [2008]. Concentraci6n de sectores intensivos en conocimiento y de alta tecnología: el caso de España, *Journal of technology management & Innovation*, 3(4), pp. 66–79.

Geels, F. W. [2004]. From sectoral systems of innovation to socio-technical systems: Insights about dynamics and change from sociology and institutional theory, *Research policy*, 33(6), pp. 897–920.

Grandori, A., and Kogut, B. [2002]. Dialogue on organization and knowledge, *Organization Science*, 13(3), pp. 224–231.

Grimpe, C., and Kaiser, U. [2010]. Balancing internal and external knowledge acquisition: The gains and pains from R&D outsourcing, *Journal of Management Studies*, 47(8), pp. 1483–1509.

Groznik, A., and Maslaric, M. [2010]. Achieving competitive supply chain through business process re-engineering: A case from developing country, *African Journal of Business Management*, 4(2), p. 140.

Hagedoorn, J. [2002]. Inter-firm R&D partnerships: An overview of major trends and patterns since 1960, *Research Policy*, 31(4), pp. 477–492.

Hakansson, H., and Eriksson, A. K. [1993]. Getting innovations out of the supplier networks, *Journal of Business-to-Business Marketing*, 1(3), pp. 3–34.

Hamel, G., and Prahalad, C. K. [1994]. *Competing for the Future* (Harvard Business Press Cambridge, MA, USA).

Hansen, P. A., and Serin, G. [1997]. Will low technology products disappear?: The hidden innovation processes in low technology industries, *Technological Forecasting and Social Change*, 55(2), pp. 179–191.

Harhoff, D., Henkel, J., and von Hippel, E. [2003]. Profiting from voluntary information spillovers: How users benefit by freely revealing their innovations, *Research Policy*, 32(10), pp. 1753–1769.

Harrison, S. H., and Corley, K. G. [2011]. Clean climbing, carabiners, and cultural cultivation: Developing an open-systems perspective of culture, *Organization Science*, 22(2), pp. 391–412.

Hatch, N. W., and Dyer, J. H. [2004]. Human capital and learning as a source of sustainable competitive advantage, *Strategic Management Journal*, 25(12), pp. 1155–1178.

Henderson, R., and Cockburn, I. [1996]. Scale, scope, and spillovers: The determinants of research productivity in drug discovery, *The RAND Journal of Economics*, 27(1), pp. 32–59.

Hendrick, T. E., and Ellram, L. M. [1993]. *Strategic Supplier Partnering: An International Study*, Center for Advanced Purchasing Studies.

Herstatt, C., and von Hippel, E. [1992]. From experience: Developing new product concepts via the lead user method: A case study in a "low-tech" field, *Journal of Product Innovation Management*, 9(3), pp. 213–221.

Hirsch-Kreinsen, H., Jacobson, D., Laestadius, S., and Smith, K. H. [2005], *Low and Medium Technology Industries in the Knowledge Economy* (Peter Lang, Berlin).

Hitt, M. A., Ireland, R. D., and Lee, H. [2000]. Technological learning, knowledge management, firm growth and performance: An introductory essay, *Journal of Engineering and Technology Management*, 17(3), pp. 231–246.

Homburg, C., Wieseke, J., and Bornemann, T. [2009]. Implementing the marketing concept at the employee–customer interface: The role of customer need knowledge, *Journal of Marketing*, 73(4), pp. 64–81.

Howells, J. R. [2002]. Tacit knowledge, innovation and economic geography, *Urban Studies*, 39(5–6), pp. 871–884.

Huang, Y.-A., Chung, H.-J., and Lin, C. [2009]. R&D sourcing strategies: Determinants and consequences, *Technovation*, 29(3), pp. 155–169.

Huber, G. P. [1991]. Organizational learning: The contributing processes and the literatures, *Organization Science*, 2(1), pp. 88–115.

Imai, M. [1986]. *Kaizen*, Vol. 201 (Random House Business Division, New York).

Jansen, J. J. P., Van Den Bosch, F. A. J., and Volberda, H. W. [2006]. Exploratory innovation, exploitative innovation, and performance: Effects of organizational antecedents and environmental moderators, *Management Science*, 52(11), pp. 1661–1674.

Janssen, O. [2000]. Job demands, perceptions of effort-reward fairness and innovative work behavior, *Journal of Occupational and Organizational Psychology*, 73(3), pp. 287–302.

Janssen, O. [2003]. Innovative behaviour and job involvement at the price of conflict and less satisfactory relations with co-workers, *Journal of Occupational And Organizational Psychology*, 76(3), pp. 347–364.

Jeppesen, L. B., and Frederiksen, L. [2006]. Why do users contribute to firm-hosted user communities? The case of computer-controlled music instruments, *Organization Science*, 17(1), pp. 45–63.

Kang, S. C., Morris, S. S., and Snell, S. A. [2007]. Relational archetypes, organizational learning, and value creation: Extending the human resource architecture, *Academy of Management Review*, 32(1), pp. 236–256.

Kanter, R. [1985]. Supporting innovation and venture development in established companies, *Journal of Business Venturing*, 1(1), pp. 47–60.

Kanter, R. M. (1988a). Three tiers for innovation research, *Communication Research*, 15(5), pp. 509–523.

Kanter, R. M. (1988b). *When a Thousand Flowers Bloom: Structural, Collective, And Social Conditions For Innovation in Organization* (Harvard University, Boston).

Katz, M. L., and Shapiro, C. [1985]. On the licensing of innovations, *The RAND Journal of Economics*, 16(4), pp. 504–520.

Knudsen, P., and Mortensen, T. [2011]. Some immediate — but negative — effects of openness on product development performance, *Technovation*, 31(1), pp. 54–64.

Kogut, B., and Zander, U. [1992]. Knowledge of the firm, combinative capabilities, and the replication of technology, *Organization Science*, 3(3), pp. 383–397.

Krippendorff, K. [1989]. On the essential contexts of artifacts or on the proposition that "design is making sense (of things)", *Design Issues*, 5(2), pp. 9–39.

Laursen, K., and Salter, A. [2004]. Searching high and low: What types of firms use universities as a source of innovation? *Research Policy*, 33(8), pp. 1201–1215.

Lin, C., Tan, B., and Chang, S. [2002]. The critical factors for technology absorptive capacity, *Industrial Management & Data Systems*, 102(6), pp. 300–308.

Magnusson, P. R. [2009]. Exploring the contributions of involving ordinary users in ideation of technology-based services, *Journal of Product Innovation Management*, 26(5), pp. 578–593.

Mahr, D., and Lievens, A. [2012]. Virtual lead user communities: Drivers of knowledge creation for innovation, *Research Policy*, 41(1), pp. 167–177.

Marsili, O., and Salter, A. [2006]. The dark matter of innovation: Design and innovative performance in Dutch manufacturing, *Technology Analysis & Strategic Management*, 18(5), pp. 515–534.

Martin, X., and Salomon, R. [2003]. Knowledge transfer capacity and its implications for the theory of the multinational corporation, *Journal of International Business Studies*, 34(4), pp. 356–373.

McCutcheon, D. M., Grant, R. A., and Hartley, J. [1997]. Determinants of new product designers' satisfaction with suppliers' contributions, *Journal of Engineering and Technology Management*, 14(3), pp. 273–290.

Melkas, H., and Harmaakorpi, V. [2011]. *Practice-based Innovation: Insights, Applications and Policy Implications* (Springer Science & Business Media, Berlin, Heidelberg, Germany).

Mintzberg, H. [1979]. *The Structuring of Organizations: A Synthesis of the Research* (Prentice-Hall, Englewood Cliffs, N.J.).

Morgan, J. [2014]. *The Future of Work: Attract New Talent, Build Better Leaders, and Create a Competitive Organization* (John Wiley & Sons, Hoboken, New Jersey, USA).

Morrison, P. D., Roberts, J. H., and Midgley, D. F. [2004]. The nature of lead users and measurement of leading edge status, *Research Policy*, 33(2), pp. 351–362.

Mowery, D. C., Oxley, J. E., and Silverman, B. S. [1998]. Technological overlap and interfirm cooperation: Implications for the resource-based view of the firm, *Research Policy*, 27(5), pp. 507–523.

Nonaka, I., and Takeuchi, H. [1995]. *The Knowledge Creation Company: How Japanese Companies Create the Dynamics of Innovation* (Oxford University Press, Oxford).

Nuvolari, A. [2004]. Collective invention during the British Industrial Revolution: The case of the Cornish pumping engine, *Cambridge Journal of Economics*, 28(3), pp. 347–363.

Nye, D. E. [2007]. *Technology Matters: Questions to Live With* (MIT Press, USA).

Pavitt, K. [1984]. Sectoral patterns of technical change: Towards a taxonomy and a theory, *Research Policy*, 13(6), pp. 343–373.

Petit, M. L., and Tolwinski, B. [1999]. R&D cooperation or competition? *European Economic Review*, 43(1), pp. 185–208.

Poetz, M. K., and Schreier, M. [2012]. The value of crowdsourcing: Can users really compete with professionals in generating new product ideas? *Journal of Product Innovation Management*, 29(2), pp. 245–256.

Polanyi, M. [1966]. The logic of tacit inference, *Philosophy*, 41(155), pp. 1–18.

Raymond, E. [1999]. The cathedral and the bazaar, *Philosophy & Technology*, 12(3), p. 23.

Robertson, T. S., and Gatignon, H. [1998]. Technology development mode: A transaction cost conceptualization, *Strategic Management Journal*, 19(6), pp. 515–531.

Rosenberg, N. [1982]. *Inside The Black Box: Technology and Economics* (Cambridge University Press, USA).

Schiele, H. [2006]. How to distinguish innovative suppliers? Identifying innovative suppliers as new task for purchasing, *Industrial Marketing Management*, 35(8), pp. 925–935.

Schweisfurth, T. G. [2017]. Comparing internal and external lead users as sources of innovation, *Research Policy*, 46(1), pp. 238–248.

Sivadas, E., and Dwyer, F. R. [2000]. An examination of organizational factors influencing new product success in internal and alliance-based processes, *Journal of Marketing*, 64(1), pp. 31–49.

Steensma, H. K., and Corley, K. G. [2001]. Organizational context as a moderator of theories on firm boundaries for technology sourcing, *Academy of Management Journal*, 44(2), pp. 271–291.

Teece, D. J. [1996]. Firm organization, industrial structure, and technological innovation, *Journal of Economic Behavior & Organization*, 31(2), pp. 193–224.

Tether, B. S. [2002]. Who co-operates for innovation, and why: An empirical analysis, *Research Policy*, 31(6), pp. 947–967.

Tidd, J., and Bessant, J. [2014]. *Strategic Innovation Management*, 1st edn. (John Wiley & Sons Inc, Hoboken).

Unsworth, K. [2001]. Unpacking creativity, *Academy of Management Review*, 26(2), pp. 289–297.

Urban, G. L., and von Hippel, E. [1988]. Lead user analyses for the development of new industrial products, *Management Science*, 34(5), pp. 569–582.

Van de Ven, A. H. [1986]. Central problems in the management of innovation, *Management Science*, 32(5), pp. 590–607.

Vega-Jurado, J., Gutiérrez-Gracia, A., Fernández-de-Lucio, I., and Manjarrés-Henríquez, L. [2008]. The effect of external and internal factors on firms' product innovation, *Research Policy*, 37(4), pp. 616–632.

Verganti, R. [2008]. Design, meanings, and radical innovation: A metamodel and a research agenda, *Journal of Product Innovation Management*, 25(5), pp. 436–456.

Verona, G., and Ravasi, D. [2003]. Unbundling dynamic capabilities: An exploratory study of continuous product innovation, *Industrial and Corporate Change*, 12(3), pp. 577–606.

von Hippel, E. [1976]. The dominant role of users in the scientific instrument innovation process, *Research Policy*, 5(3), pp. 212–239.

von Hippel, E. [1986]. Lead users: A source of novel product concepts, *Management Science*, 32(7), pp. 791–805.

von Hippel, E. [2005]. *Democratizing Innovation* (MIT press, USA).

von Hippel, E. [2009]. Democratizing innovation: The evolving phenomenon of user innovation, *International Journal of Innovation Science*, 1(1), pp. 29–40.

Wadell, Ölundh, Björk, and Magnusson. [2013]. Exploring the incorporation of users in an innovating business unit, *International Journal of Technology Management*, 61(3/4), pp. 293–308.

Walsh, V. [1996]. Design, innovation and the boundaries of the firm, *Research Policy*, 25(4), pp. 509–529.

Williamson, O. E. [1985]. *The Economic Institutions of Capitalism: Firms, Markets, Relational Contracting*, Vol. 866 (Free Press, New York).

Zahra, S. A., Ireland, R. D., and Hitt, M. A. [2000]. International expansion by new venture firms: International diversity, mode of market entry, technological learning, and performance, *Academy of Management Journal*, 43(5), pp. 925–950.

CHAPTER 4

STRATEGIC MANAGEMENT OF TECHNOLOGY AND INNOVATION

4.1 COMPETITIVE ADVANTAGE AND TECHNOLOGY

We start with the assumption that the company's strategy focuses on a series of choices, activities, and practices aimed at providing value to the market and customers. Numerous definitions of what we mean by strategy have been provided in the literature. The explanation for the existence of different views for the term strategy over time is that strategy focuses on how to compete in the market, and this has also changed substantially over time. Thus, Alfred D. Chandler, Jr., author of *Strategy and Structure* who studied the relationship of an organization's structure and its strategy, defined strategy as "the determination of the basic long-term goals and objectives of an enterprise, and the adoption of courses of action and the allocation of resources for carrying out these goals" [Chandler, 1962]. Some other authors have approached strategy and strategic change as a way to defend the company's position in the market [Abernethy and Brownell, 1999; Miles *et al.*, 1978; Shortell and Zajac, 1990]. There are also authors such as Gebauer *et al.* [2012] and Schlegelmilch *et al.* [2003] who argue that strategic innovation is focused on customers, new market creation, and reshaping customer needs. Drucker [1994] defines strategy as the manager's plan on how to gain and sustain a competitive advantage. The definition of strategy has been summarized with the work of Andrews [1971, p. 18], who defined strategy as the "pattern of objectives, purposes, or goals and the major policies and plans for achieving these goals, stated in such a way as to define what business the company is in or is to be in and the kind of company it is or is to be." However, if there is a referent in the world of strategy, it is Michel Porter, Harvard Professor, who became one of the cornerstones in the strategy realm and the main reference for strategic thinking with the publication of his book, *Competitive Strategy*. Professor Porter defined competitive strategy as "a broad formula for how a business is going to compete, what its goals should be, and what policies will be needed to carry out those goals" [Porter, 1985]. This vision is oriented to the detection of core resources, capabilities, and competencies that are critical to compete, showing strategy as a commitment to undertake a series of some actions rather than others, involving at the same time a series of resources [Oster, 1999; Grant, 2002]. In conclusion, we can say that strategy tries to answer two questions: Why do some companies triumph while others fail? And what can managers do about it?

Professor Porter explains the way in which businesses compete through the concept of competitive advantage. Thus, Porter states, "Competitive advantage grows out of value a firm is able to create for its buyers that exceeds the firm's cost of creating it. Value is what buyers are willing to pay, and superior value stems from offering lower prices than competitors for equivalent benefits or providing unique benefits that more than offset a higher price. There are two basic types of competitive advantage: cost leadership and differentiation" [Porter, 1985, p. 5].

Therefore, Porter recognized the existence of two types of competitive advantages: differentiation and cost leadership. A company holds a competitive advantage when the market recognizes a higher value in that company's products than in those of the competition. In this way, the company's strategy is oriented either toward generating differentiated products and services that are valued by its customers or to achieving leadership in its costs that increases its margins or has an impact on its prices. Be that as it may, the generation of competitive advantage is supported by a series of elements that Porter analyses through the value chain of the company.

Three elements — the perceived value by customers, the price, and the cost of production — intervene in the game of generation of competitive advantages. Accordingly, the greater the difference between the perceived value and price, the greater the probability that a company competes with guarantees in the market. In the same way, the greater the difference between their prices and cost structure, the greater the margin that is recognized by the market. Thus, a company will have a competitive

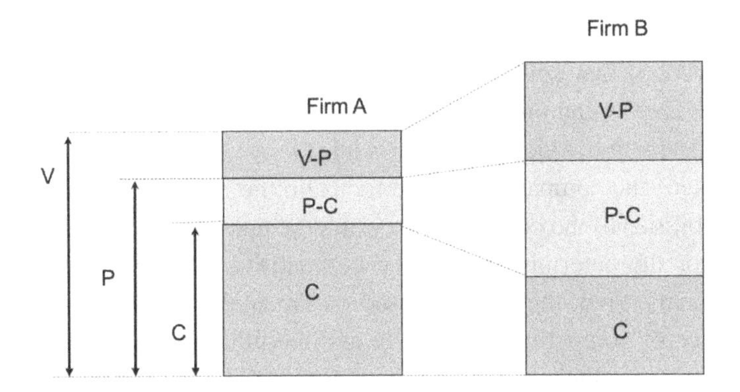

Figure 4.1. The Role of Value Creation and Cost in the Generation of Competitive Advantage.

Source: Rothaermel (2008).

advantage over another competitor if the value attributed by the customer is greater and its margins are greater (Fig. 4.1).

Some authors, like Grant in his book *Contemporary Strategy Analysis*, identify competitive advantage with profitability; thus, he states, "when two or more firms compete within the same market, one firm possesses a competitive advantage over its rivals when it earns (or has the potential to earn) a persistently higher rate of profit," and adds, "In the long run, competition eliminates differences in profitability between competing firms, hence, competitive advantage is a disequilibrium phenomenon that is a consequence of change" [Grant, 2002, p. 211].

Therefore, being aware of the aforementioned, it would be interesting to identify the different sources of change in order to understand how firms can harness emerging opportunities to build a competitive advantage. The company may face different changes that can be a threat or an opportunity depending on its ability to take advantage of them. These changes may be due to geopolitical, demographic, or social factors, and economic, environmental, legal, or technological issues. The company itself can also be an important and valuable source of change, since, based on its resources and capabilities, it can implement strategies aimed at building competitive advantages. Much of the research that has been carried out in this area has focused either on the analysis of the opportunities and threats that exist in the environment and the market [Porter, 1980, 1985] or on the analysis of the capacities and resources of the company [Hofer and Schendel, 1978; Penrose, 2013] that could be used to respond to the different challenges that arise. Depending on whether we focus on the analysis of the environment to determine which market conditions can enhance the competitive advantage of the company, or if we focus on the analysis of the resources and capabilities of the company, we would be talking about environmental models for the determination of the competitive advantage or resource-based models. First, the models based on the analysis of the environment as a source of competitive advantage assumed that the resources used by the companies in the same market were not differential and that when differences occurred, these were corrected quickly due to the availability and mobility of resources between companies. It seems that this option does not reflect an approach in line with reality, since the appropriability of a given class of resources, such as technology, by a company can be a differentiating factor against competition and the source of a competitive advantage. Hence, the resource-based approach seems to be more useful in understanding how to build competitive advantages.

According to Daft [2010], company resources include all the assets of a company, its capabilities, organizational processes, company attributes, and information and knowledge, which are controlled by the company and through which they can implement strategies for their effectiveness and efficiency. In the strategic language, we can consider some resources as strengths that will provide the firm with the basis for the construction of competitive advantages. In order to consider a resource as a strength for the company, it must go beyond being a mere asset necessary for the performance of the business activity and must also be constituted as an exclusive resource for the company itself and, therefore, be able to differentiate it from the market.

Following Barney [1991], we distinguish between three types of resources: physical capital resources, human capital resources, and organizational capital resources. Physical capital resources include the technology used by the firm, the plants and equipment and access to raw materials [Williamson, 1983]. Human capital resources, in line with the contributions of Becker [1964], are composed of training, experience, and relationships. Finally, organizational capital resources are defined as the formal reporting structure of the firm, formal and informal planning, controlling, etc. Therefore, we can observe how, on the one hand we have the physical resources of the company and on the other hand the way in which the organization manages itself and organizes its resources through the people's capabilities, organizational structure, and company culture.

Therefore, technology can be considered a resource of the company in which it can base its competitive advantage and can provide a basis for the differentiation or efficiency of the company. However, technology itself could not be enough to build up a solid and consistent competitive advantage, since human capital capabilities, skills, and knowledge are necessary to implement a unique use of technology. In the same way, the company must be aware of the evolution of technology over time, since a company that has a differential technology can see how it might be replaced by another. Given the above, we face the question about the role of technology in the development of competitive advantages for the company. To answer this question, it is necessary to delve into the process of formation of competitive advantages.

Michael Porter, in his work *Competitive Advantage* refers to the role of technology in the process of constituting and forming competitive advantages. Thus, Porter makes reference to all the technologies of the company, emphasizing not only those applied to the product (goods or services) but all those that are being used anywhere in the value chain of the company. Porter enhances the role of technology in the firm by affirming that, "A firm as a collection of activities, is a collection of technologies" [Porter, 1985, p. 166]. In this sense, technologies are present not only in primary activities of the value chain but also in all support activities.

For a better understanding of the above, we are going to go deeper into the definition of the value chain. The value chain represents all the subprocesses of the company through which it contributes value to the products and services offered to the market. This is carried out not only through core or primary processes but also through certain support processes. According to Porter [1985], on the one hand, primary activities include the following:

- *Inbound Logistics*: In this section, we include all activities related to suppliers, storage, and availability of goods in the production process.

- *Operations*: They include all activities aimed at transforming the different inputs into goods and services. The activities of production (goods) and operations (services) are included in this section.
- *Outbound Logistics*: These include all the activities of storage, transportation, and distribution of the finished products of the company.
- *Marketing and Sales*: We include all systems that communicate the value proposition of products and services of the company to consumers.
- *Service*: It includes all the activities that guarantee the correct operation over time of the goods and services of the company.

On the other hand, secondary activities is comprised of the following:

- *Procurement*: These activities include all the processes of acquisition of inputs and resources necessary for the activities of the company.
- *Human Resource Management*: This consists of all activities of recruitment, hiring, training, development, motivation, reward systems, and human capital retention.
- *Technological Development*: It relates to the acquisition and development of knowledge necessary for the company, including the protection of the company's knowledge base, staying current with technological advances, and maintaining the technological expertise necessary for value creation.
- *Infrastructure*: It includes the organizational structure and its management, administration, control, and quality.

To illustrate the above, we find it interesting to give an overview of the main technologies used in the different activities of the value chain, making reference to the technologies existing at the time of the writing of this work. Starting with inbound logistics, we can refer to all the technologies related to transportation, optimization of logistics routes, automation of material logistics processes, integrated supply chain management (SCM), or traceability systems. Of particular interest in this area are radio-frequency identification technologies (RFID), which are more accessible now due to their lower costs, the use of the Internet of things (IoT), and the implementation of automatic identification and data capture technologies (AIDC). Another field of interest may be robotics, which has already been successfully applied through the use of autonomous mobile robots that replace the use of forklifts managed by human personnel which can make the processes of materials management more efficient.

Second, we want to refer to production process technologies that are central to the value chain of the company. In many ways, production has become an outsourced process based on its efficiency. Information and communication technologies and their integration with manufacturing technologies are a constant source of efficiency. Thus, we can cite, as relevant, the robotics applied to the automation of the production processes or service rendering, optimization of cross-supply demand based on intensive use of data, reduced time to market, intelligent processing of resources and raw materials, systems for energy efficiency, predictive and remote maintenance, man–machine collaboration, inventory management, 3D printing, distributed production, and advanced quality controls, among others.

In addition, as we have mentioned earlier in Chapter 1, nanotechnology and biotechnology will also be core technologies used to increase added value, changing many of the products and industrial processes as we know them today.

With regard to technologies linked to distribution and marketing processes, we can point to e-commerce, including B2B, as the cornerstone, resulting in advanced logistics systems for the fast delivery of all types of products. In particular, it is necessary to emphasize the technologies that allow mobile electronic commerce, turning this into a strong sales channel. All of the above would not be possible without the support of information and communication technologies, which support the advanced use of data. Finally, reference must also be made to gateways and payment processing systems, as well as technological platforms that facilitate the creation of global markets for the exchange of goods and services.

Finally, within the primary value chain processes, we can talk about customer service processes. This interaction with the customer is intensive in the use of technologies, especially through the implementation of automated customer relationship management systems (CRM). Increasingly important are the presence of multi-channels (omni-channel), the use of Big Data that allow individualization of attention, and artificial intelligence, which can replace many of the interactions that occur among humans.

Returning to the question we faced earlier, it is noteworthy to know when technology can be relevant to achieve a competitive advantage. In this sense, and as Professor Porter points out, technology influences our competitive advantage if it is capable of playing a significant role in determining our cost structure or in our ability to differentiate. In other words, technology will be able to impact our competitive advantage if it influences our cost drivers or sources of uniqueness. On the one hand, among the cost drivers, we can highlight the scale (purchasing, production, processing), learning in the firm that leads to more efficiency, linkages between different parts of the value chain, pattern of capacity utilization, integration of activities, etc. On the other hand, differentiation drivers can be built based on the uniqueness of the firm's products or services, infrastructure, superior workforce, exclusive knowledge generated through research and development and technology development, more efficient procurement or inbound logistics systems, more efficient or best quality production systems, better distribution networks, unique marketing channels, highly responsive customer services, etc.

Therefore, the company must be aware of the technologies that can be sources of competitiveness and that could be applied to build up either a cost or differentiation advantage for the firm. In this sense, it is advisable that organizations follow the technologies relevant to its activity and incorporate those that may be interesting.

4.2 COMPETITIVE DYNAMICS AND TECHNOLOGICAL INNOVATION

To understand how companies compete and how technological change can influence it, it is important to know the different theories and explanatory models of technological evolution over time. Chapter 1 of this book already referred to the different revolutions or technological phases that have occurred throughout human history. Now we are interested in understanding the factors and dynamics that influence this technological evolution. Companies carrying out their activities in a specific market are interested in knowing the emerging innovations in their area of action that can erode their competitive advantage. One of the issues managers want to know is when and how innovations that appear on the market can influence their competitive advantage. In this way, foreseeing how a technology can evolve can enable these companies to respond more efficiently in the market, even by defending themselves against the possibility of new entrants.

One of the first theories to which we must refer to explain the evolution of technology is the dominant design theory, which can be attributed to authors such as Abernathy and Utterback [1978], who studied the patterns of industrial innovation in the 1970s. This theory defends that in the early stages of market evolution, which is characterized by technological uncertainty, numerous designs emerge and coexist. At a given moment, the market prioritizes a certain product design, which entails a particular technological solution. We can define product design as the set of specifications or product characteristics that determine the predominant structure of the product category [Christensen *et al.*, 1998]. The dominant design does not imply technological superiority, but the solution facilitated is the one that best combines the technological possibilities and commercial needs and opportunities [Tushman and Rosenkopf, 1992; Wade, 1995]. When a dominant design appears, suppliers and other actors around that product have a framework to direct their efforts of improvement and technological evolution. Innovative effort toward product design alternatives declines (product technology), while focusing on improving and evolving dominant design architecture (process technology) (Fig. 4.2).

We make a distinction between the concept of dominant design and that of standards. Standards, from an engineering point of view, can be defined as the set of quality, compatibility, or connectivity specifications that are required for the product to function properly [Grindley, 1995]. The main difference between a dominant design and a standard is that the latter basically serves a functional feature, while the former must include market acceptance as a necessary condition. Sometimes a dominant design can arise from the choice between different standards, as it did in the video market, where the VHS and Beta standards competed until VHS became the dominant and accepted standard design. Dominant design theories have been applied in a practical way in the analysis of sectors such as

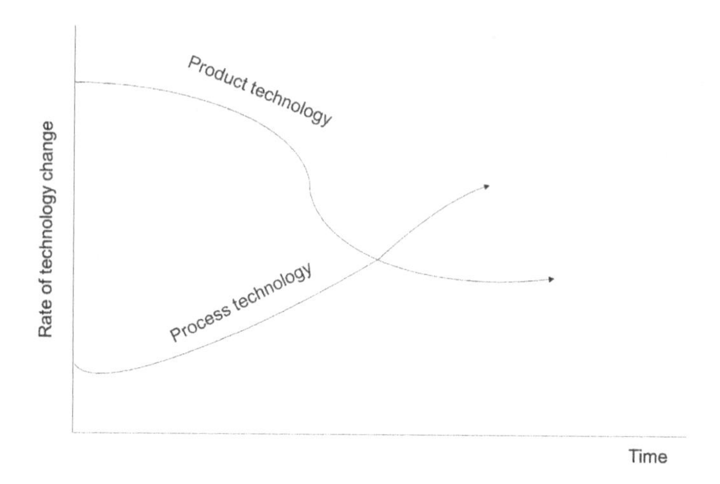

Figure 4.2. Impact of Dominant Design on the Pace of Technology Change and Processes.

Source: Abernathy and Utterback [1978, p. 2].

typewriters, televisions, electronic calculators, and of course, automobiles [Utterback, 1996].

One of the references to understanding technological evolution is found in what is known as the technological S-curve [Foster, 1986]. The S-curve is intended to predict when a new technology is likely to arise. Through the study of the curve drawn by the evolution of a technology over time, we can determine the probability of the appearance of a new technology that replaces the existing, predominant technology. The reasoning is that when a technology reaches its limit of development, it is to say that when the performance of that technology no longer increases, the market starts to look for alternative technologies that could overcome the existing limits. The latter is explained by Foster [1986, p. 34]: "If you are at the limit, no matter how hard you try you cannot make progress. As you approach limits, the cost of making progress accelerates dramatically. Therefore, knowing the limit is crucial for a company if it is to anticipate change or at least stop pouring money into something that can't be improved. The problem for most companies is that they never know their limits. They do not systematically seek the one beacon in the night storm that will tell them just how far they can improve their products and processes."

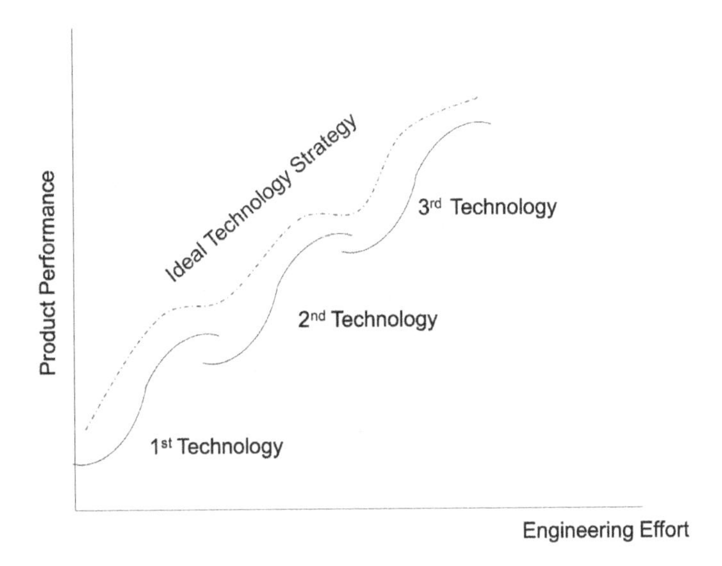

Figure 4.3. A View of Technology Strategy: Switching S-curves.
Source: Christensen [1992].

Figure 4.3 represents the well-known technological curve or S-curve. The technological curve is represented in two axes: the vertical axis represents the development achieved by this technology, while the horizontal axis measures the effort of technological development over time. The curve is characterized by a slow evolution of technological development at the beginning, which is reasonable since it is at the beginning when the technology is more incipient and the yields are more uncertain. Then there is a turning point where technology reaches a degree of development where the yields are more than proportional to the development effort of the technology. This is the stage where the performance of the technology grows faster and therefore becomes more attractive to the market of reference. However, there is a weak point in the rate of progress of technology yields. Yields grow less than proportionally to the development effort made in that technology, increasing the cost of each marginal improvement. The pace of technological progress stagnates and technology does not evolve significantly, to the point where the curve flattens out. When there are symptoms that the pace of technological evolution is softening, the players present in the market begin to consider other emerging technologies as possible substitutes for the dominant technology. As we can see from the graph, at the beginning, emerging technologies perform much lower than dominant technologies in the market. However, their evolution over time will narrow the performance differences until the two curves intersect at a given time. From that point on, the new technology is more attractive than the existing one and there is a massive migration toward it, with the previous one progressively discarded. The question that managers face is when the company should decide to embrace the new technology and abandon the existing one.

In certain cases, a technology never reaches its maturity period since there may be so-called technological discontinuities. A technological discontinuity occurs when a new technology breaks out in the market covering the same needs as an existing technology, but the latter does so with a totally different knowledge base [Anderson and Tushman, 1990]. Initially, the technological discontinuity may have yields below the established technologies, so the incumbents may be reluctant to adopt this new technology. However, if the new technology shows a greater slope, that is, if the increase in its returns is higher than the established technology, many companies can and will migrate to the new one. On the other hand, the new entrants in the market can directly adopt the new technology since it is foreseen as a greater evolution of it.

The S-curve has a number of limitations as a predictor of technological evolution, since it is really difficult to explain when it is convenient to invest in a new technology and abandon the current technology, to know how effective the new technology is when compared to the current one, or the role of the managers in the adoption and exploitation of the new technology.

Another interesting reference is the so-called "Technology and Market Trajectories Theory." This theory seeks to explain not only how technological progress occurs but also by reference to the rates of technological advancement that consumers are able to absorb. There is a highly relevant factor in the understanding of this theory: the patterns of innovation are highly influenced by the crossing of the trajectories of technological evolution and those that represent the needs of the customers.

The first references to technological trajectories as a way of representing technological progress can be found in works by authors such as Dosi [1982],

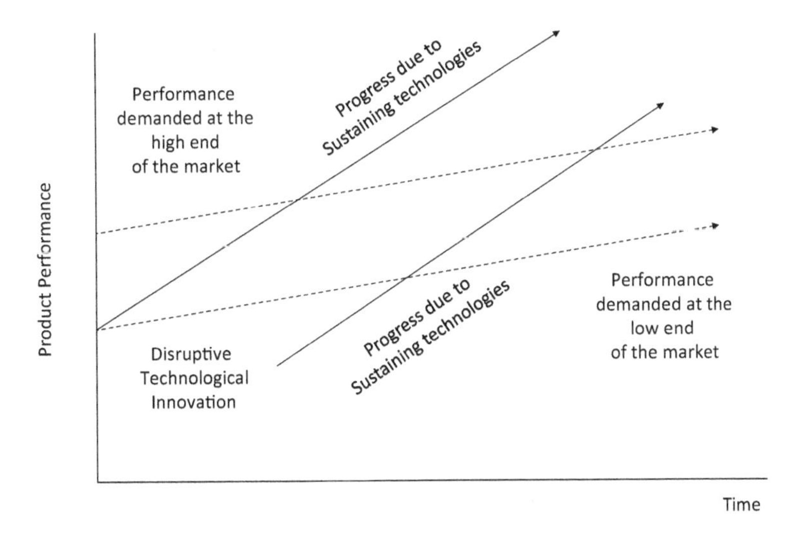

Figure 4.4. Technology Trajectories.

Source: Christensen [1997, p. xvi].

who initiated a stream of research based on this concept and applied it to different sectors [Christensen, 1993; Christensen *et al.*, 1998]. It is commonly accepted that the evolution of technological trajectory exceeds the technological demand made by consumers. This means that a technology that fails to meet the needs of consumers at first can exceed their demand, taking into account the needs of consumers in the future (Fig. 4.4).

Many technologies have a sustained impact over time as they systematically increase the performance of this technology in the market. Sometimes disruptive technologies arise offering better performance with a lower cost structure, that is, more efficiency. When this occurs, it can facilitate the appearance of new market segments attracted by these new features of emerging technologies. When the new disruptive technology is already established in the upper levels of the market, it can attack the lower levels of the market, having as incentive larger market segments and a possible greater profitability. Through the various investigations carried out, it has been concluded that companies established in the market usually choose to introduce innovations in a sustainable way, even if they come from more disruptive technologies. However, new entrants to the market are more likely to introduce disruptive technologies into the market in a new way or by targeting markets where current technologies do not appear commercially viable. These patterns of innovation have been observed in different industries such as commerce, telecommunications, semiconductors, and printing. A special reference should be made to the research carried out by Clayton Christensen, published in his book *The Innovator's Dilemma* [1997], which refers to the fact of disruptive technological change and how to explain the causes of failure of well-established companies when confronting this change. It applies to sectors such as steel, industrial excavators, photocopiers, motorcycles, computers, and management software.

4.3 DIFFUSION AND ADOPTION OF INNOVATIONS

Once we have studied, as in the previous section, the way in which the evolution of technology takes place, it is interesting to understand how the process of diffusion and technological adoption takes place. Understanding the processes of diffusion and adoption of a new technology, we will be able to make strategic decisions at the technology level. We can say that the more a new technology is adopted by consumers in a market, the more will be the diffusion of that new technology. Therefore, diffusion and adoption are two closely related but distinct processes. Next, we will explain in detail both processes in order to understand the interrelations that these have in the definition of the technological strategy.

Following Katz *et al.* [1963], we can define diffusion as a process involving acceptance, over time, of some specific item, an idea, or practice by individuals, groups, or other adopting units linked to specific channels of communication, to a social structure, and to a given system of value or culture [Katz *et al.*, 1963]. In this sense and referring to the earlier definition, the diffusion of an innovation implies not only the transfer of the innovation but also an active attitude from the receiver, that is to say, the new user of this technology.

Rogers [1962, p. 5] also offers a definition of diffusion as "the process by which an innovation is communicated through certain channels over time among the members of a social system." In this definition, Rogers emphasizes the communicative process of the term diffusion, which is the process of communication in both directions. This communication process is focused on the new idea's novelty, which is central to the communication message within the diffusion process. In this context, the degree of novelty will be related to the uncertainty assumed by the diffusion and modulating that uncertainty with the information offered will be possible, which will allow customers to choose between several possible alternatives.

Rogers also refers to the social change that occurs through the diffusion process because when a new technology alters the current state of things, it leads to social consequences. Rogers also makes reference to the difference between the terms diffusion and dissemination, where diffusion refers to a more spontaneous process of communication of new ideas and the term dissemination is more oriented toward the regulated and managed processes of communication of the novelty.

Coming back to the definition of diffusion by Rogers, this includes several elements. The first one is innovation, which is understood as "an idea, practice, or object that is perceived as new by an individual or other unit of adoption" [Rogers, 1962, p.11] and where the degree of novelty of that idea is important and is expressed in terms of knowledge, persuasion, or decision to adopt it. The second concept is communication as a process in which several participants exchange information for the purpose of common understanding. In this sense, diffusion is a communication process that refers to a new idea. The communication process is also organized through different channels of communication. Finally, the definition of diffusion also entails a social structure that defines the patterns that allow us to understand the behavior of participants.

In the scientific literature, the first references to the interpretation of innovation diffusion can be found in the work of Mansfield in which reference is made to the concept of the diffusion curve [Mansfield, 1961].

The diffusion curve can be understood as a function that explains the number of potential users who will adopt an innovation over time [Jensen, 1982]. Normally, the diffusion curve explains the diffusion pattern of a given technology over time. The percentage of adoption over time is an incremental function that can at first show a convex shape, but over time describes a concave shape, drawing a bell-shaped curve [Rogers, 1962].

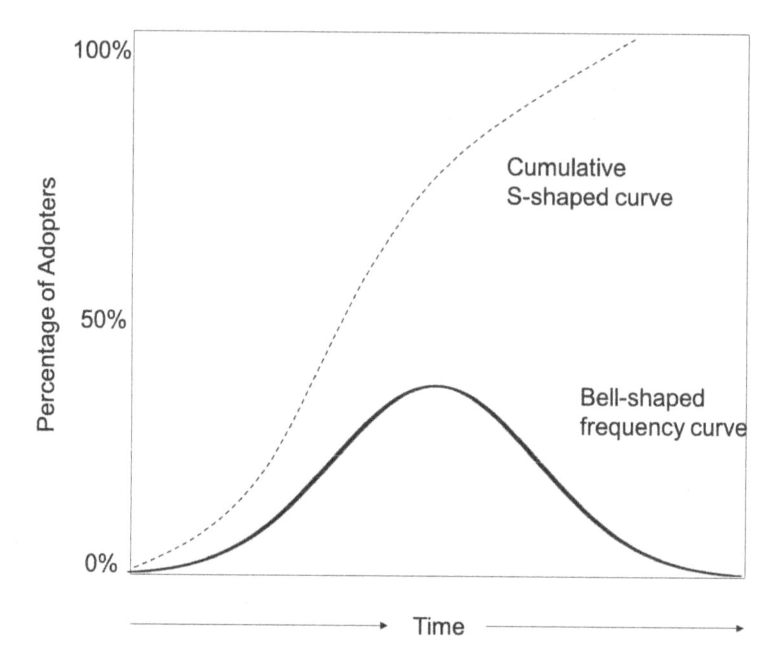

Figure 4.5. The Diffusion Curve.
Source: Rogers [1962].

As we can see in Fig. 4.5, the adoption rate begins slowly and then accelerates its pace, increasing the number of individuals in the system adopting the new technology. The adoption rate is maintained until it moderates its slope and reaches a maximum, and then decreases gradually. Rogers argues that to understand the evolution of the diffusion curve, it is necessary to determine the categories in which we can structure the adopters of this technology and determine the methods that explain in which category each member of the system can be included.

Rogers presents five different categories of adopters which are as follows (Fig. 4.6):

- *Innovators*: Innovators are characterized by the desire to try new things. This type of adopter is focused on the sources of diffusion of the innovations. They must be able to assume high risk and uncertainty of the innovations from a possible financial loss. The innovator plays an important role in the process of diffusion of innovations within the

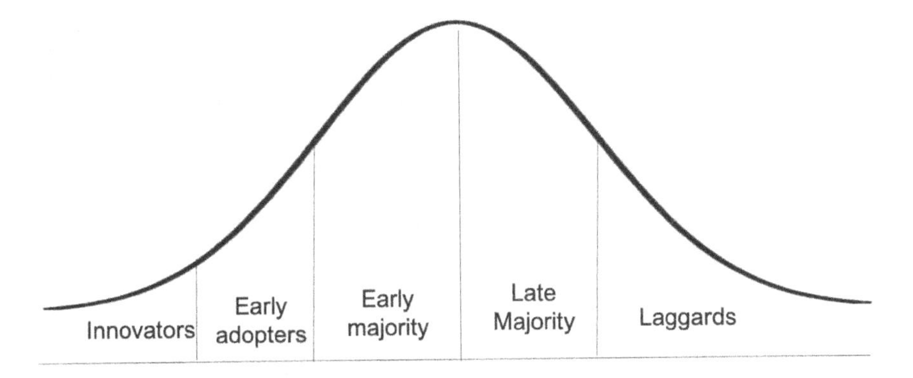

Figure 4.6. Adopter Categorization on the Basis of Innovativeness.
Source: Rogers [1962, p. 247].

social structure, since they play the role of testing and dissemination of good practices.

- *Early Adopters*: Early adopters are characterized by being integrated into the system and having the capacity to become opinion leaders. Many members of the system observe early adopters before using an innovation, while addressing them in search of advice and opinion. This category is highly respected by the members of the system, knowing that to maintain that role they must make decisions reasonably. Early adopters play an important role in reducing uncertainty within the system.
- *Early Majority*: The early majority adopts innovations just before the system average. While individuals in the system interact with them, they hardly take leadership positions. However, they do play an important role as interconnections between the most innovative groups and the followers of the system. This group has a decision period when adopting a new technology substantially greater than the previous ones; they want to be neither the first nor the last in the adoption process.
- *Late Majority*: The late majority adopts innovations just after the system average, responding to economic difficulties or system pressures. Such individuals are generally skeptical of innovations. They need to eliminate uncertainty as much as possible due to the scarcity of resources.

- *Laggards*: This group is the last in adopting innovations. Normally, when laggards have adopted an innovation, there is already a new technology in the market that has overtaken the former one. They are really suspicious about the role of innovations. They focus a lot on the past instead of looking to the future and have an economic position that does not allow them to take risks in adoption in the face of possible economic loss.

The interest of researchers has been on focused on studying the speed at which diffusion occurs and on what factors the delay in adopting them depends [Mansfield, 1968]. The emergence of innovations creates imbalances in function of the new technological dynamics that have to be faced by the participants in the market. Then, the process of technological adoption is a decision process in order to overcome these imbalances.

The key factor in the process of adopting a new technology by a company is the economic advantage that will generate the use of this new technology. However, there is uncertainty about what that economic advantage may be, while it is certain that the company assumes fixed costs with the incorporation of this new technology. Therefore, the adoption decision will take place when that level of uncertainty is reduced to a level acceptable to the adopter.

The level of uncertainty can be lowered through the information process. This information process can be carried out through the interaction between users or non-users [Mansfield, 1961], between suppliers and non-users, or through a trial process. These models of explanation on the diffusion of innovations have been known as learning models and have suffered criticism because they assumed that all the participants were equally susceptible to any innovation.

Thus, new models appeared, trying to explain the diffusion process as a function of the differences between companies, known as probit models. In these models, a company makes the decision to invest in a new technology based on the expected pay-offs. As expected pay-offs can vary across different companies depending on the characteristics of the firm, a probit model is able to explain the different speed and rate of adoption of the innovations in different companies [David, 1969]. Therefore, in these models, diffusion occurs following a learning process, since an adopter decides to adopt a particular technology by observing the rate of return from other incumbents in the market.

Another orientation to the process of diffusion can be observed in the Game Theoretic Approach, where the diffusion process is the result of the strategic behavior of the different participants in the market trying to decide the optimal time to adopt an innovation to get ahead of other participants. Reinganum [1981] considers both a decreasing cost in the adoption of an innovation and a decreasing return or profit to be gained with an increase in the number of users. This contrasts with the probit models where it is assumed that there is independence among the adopters. She also shows that, even if the companies are identical, the pace of adoption will vary, since it will depend on the strategic interactions between the different companies.

4.4 INNOVATION AND TECHNOLOGY STRATEGIES

There are authors such as Szakonyi [1990] who stress the importance of the linkage between technology and business strategy, identifying the shortage of R&D leverage of American companies as the cause of the lack of competitiveness in comparison with Japanese firms. In this vein, Weil and Cangemi [1983] observed this lack of linkage between research and business strategy. The authors reported a mismatch of vision between researchers and managers, lack of knowledge of corporate goals by researchers, and difficulty in making long-term forecasts about technological trends.

The main objective of a technology strategy is to guide the company in the identification, incorporation, and use of technology to achieve and sustain a competitive advantage. There are authors who defend that the determination of a technological strategy consists in the identification and selection of technological projects that the company wants to undertake [De Meyer, 2008].

In this same vein, there are authors who argue that the technological strategy is focused not only on the internal technologies of the company but also on the links with the technological systems to the environment. Thus, the technological strategy pursues the understanding and importance of the internal and external technologies that the company must use [Ford and Thomas, 1997].

Technological strategy is restricted not only to high-tech companies but also to those companies oriented to the client, or their capabilities also require a technology strategy. The technological strategy determines the choice of technical capabilities and the product and process platforms available to the company [Bone and Saxon, 2000].

Subsequently, the technological strategy must be constantly reviewed and monitored according to the trends of the environment. The inputs of this ongoing review are important in reconfiguring the company's technological capabilities [Davenport *et al.*, 2003].

Coombs and Richards [1990] analyzed R&D departments and found an increased need for managers to orientate strategically the need of products and technology portfolios and reported that R&D unrelated to corporate goals was due to a lack of effort and resources.

Studies of strategy and technology have adopted different approaches, from those based on technology foresight [Twiss, 1992; Szakonyi, 1990] to those focused on technology audit [Fusfeld, 1978; Clarke *et al.*, 1989] or technology portfolio [Coombs and Richards, 1990].

The first approach to the role of technology in strategy is what Roussel *et al.* [1991] called the first-generation R&D mode, also known as the "strategy of hope," where firms hired qualified labor, provided them with the necessary resources, and thus expected them to produce commercially viable results. In the 1950s and 1960s, the strategy of hope was possible. Subsequently, increased competition, the pause in demand, and the pressure on profitability made this approach no longer viable. This led companies to a second generation of R&D, which organized each of the projects by analyzing their costs and benefits. However, under this second approach also, the management of R&D did not have a strategic orientation, since project-by-project management led to a lack of general vision about the overall strategy.

To overcome these problems, Roussel *et al.* [1991] referred to what they called "Third-generation R&D management." Under this approach, managers take a holistic view of R&D in relation to the general activity of the company and its different units, working at the corporate level.

The world is becoming more and more competitive, ranging from competing under the parameters of cost and quality to others that establish the time to market and knowledge as reference variables. In this sense,

D'aveni [1994] defines hyper-competitiveness as characteristic of markets. It is becoming increasingly important to be aware of the need to align the company's innovation and technological strategy with its competitive strategy. Many companies focus their efforts on launching products that ultimately do not succeed in the market, while many of the market demands remain unfilled. Therefore, it is increasingly important to propose frameworks that analyze the role of technology and innovation within the overall strategy of the company. Therefore, authors such as Hax and Majluf [1984] and Hax and No [1993] link the company's overall strategy to its strategy of innovation and technological strategy.

Hax and Majluf [1984] defend the linkage of the technological strategy with the general strategy of the company as one of the functional strategies. The unit of analysis is the Strategic Technical Unit (STU), which entails all technologies, skills, and disciplines linked to a product or process in order to gain technological advantage. These authors define three levels of analysis: the corporate level, the business level, and the STU level. The process of strategic definition begins with the analysis of the existing technologies in the environment (technological intelligence, opportunities and threats in the technological field, technological attractiveness). It continues with the internal analysis of the STU, defining weaknesses and threats and distinctive technological competencies. After the analysis phase, a strategic formulation phase is established, specifying technological policies and a series of action programs. Finally, there is the strategic programming and budgeting phase (Fig. 4.7).

Chiesa and Mazini [1998] criticize this approach since, according to these authors, the approach to the product and its technologies can limit its application to a limited number of industries where the orientation of products and processes is very well defined.

Once again, it is interesting to refer to Michael Porter and his work *Competitive Advantage* [Porter, 1985] to understand the dynamics of technology-driven strategy. We have already discussed earlier in this chapter how technology can serve as a basis for the generic strategies outlined by Porter — either differentiation or cost leadership. This approach extends the vision from the product focus to the value chain of the company. Therefore, the technology extends its role to support the management and the organization of certain skills in the value chain. Either way,

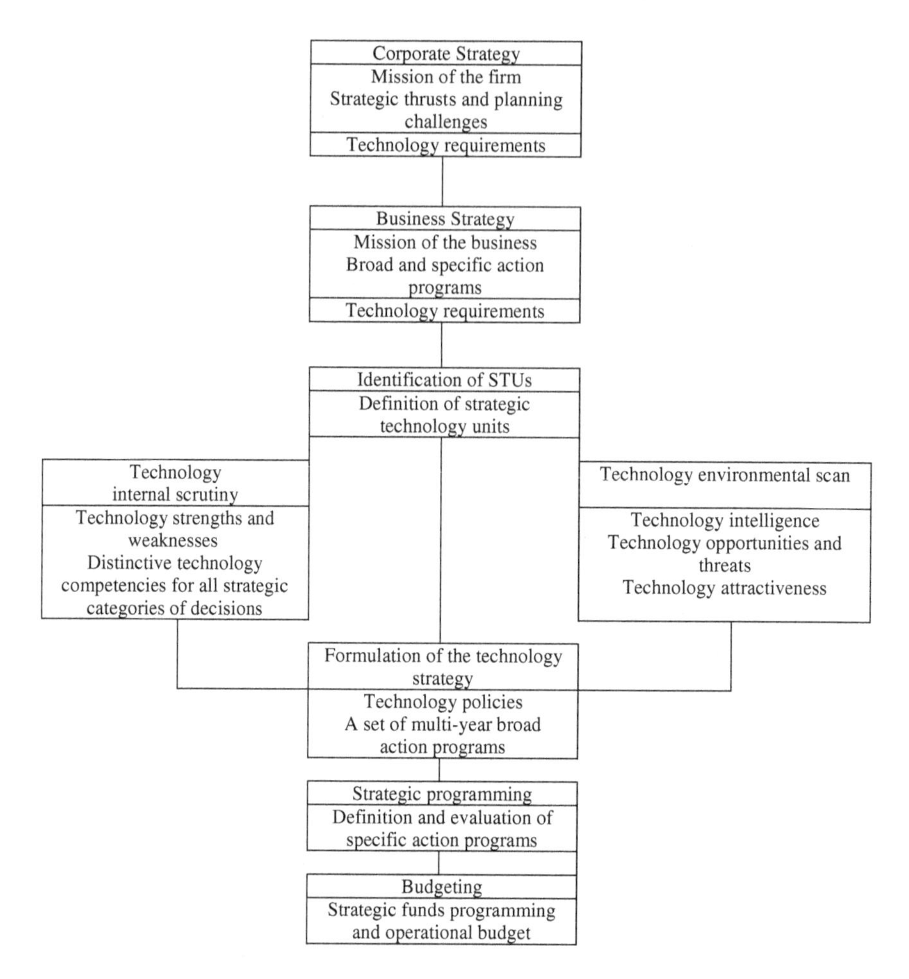

Figure 4.7. A Framework for the Development of Technology Strategy.

Source: Hax and No [1993].

Porter suggested that technological change can lead to a competitive advantage, as long as it leads to differentiation of the firm or cost leadership and can be protected from imitation. Also, technological change can be considered to change the drivers of cost or uniqueness in favor of the company.

It can also be considered that the strategy of leading technological change can lead to a number of advantages of the pioneer, in addition to those of technological change. Porter says that even if the market imitates the technologies incorporated by the pioneer, it would have been able to generate a number of advantages by being the first to move in the market. Finally, a technological change that shifts the overall industry is also desirable, even though it is easy imitable.

To understand the strategic dynamics linked to technology, we have to refer to the market itself and how technology can influence market forces. To do this, following the contributions of Porter, we will analyze the role of technology in the bargaining power of suppliers and customers, the threat of new entrants, substitute products, and rivalry between the suppliers as follows:

- Technology can be the basis for an entry barrier. Thus, the incorporation of new production technologies can generate economies of scale that hinder the arrival of new competitors. In the same way, technology can influence the learning effect of the company, affecting its layout, yields, and speed of production. The incorporation of technology may also entail cost advantages or product differentiation that may pose significant barriers to new entrants. Finally, it is also interesting to note that new technology can influence the training and necessary equipment of buyers and distribution channels.
- Technology can also influence the bargaining power of both suppliers and customers. In this sense, it is clear that technology can change the relationship between the industry and its buyers or suppliers. In the particular case of buyers, technology can directly affect their power of backward integration. As far as suppliers are concerned, technology can force the industry to buy from a supplier because of the technology it uses. The investment in technology by the company can eliminate the dependence of a single supplier or the creation of internal knowledge that allows it to extend the range of suppliers to work with.
- Technology is key to the appearance of substitute products. In fact, the appearance of substitute products will depend on the value function of these products in relation to their price and the costs associated with the change. Technology change creates entirely new value products and functions that can be attractive to the buyer.
- Finally, technology clearly influences the degree of rivalry of an industry, since it affects its cost structure or the capacity of differentiation

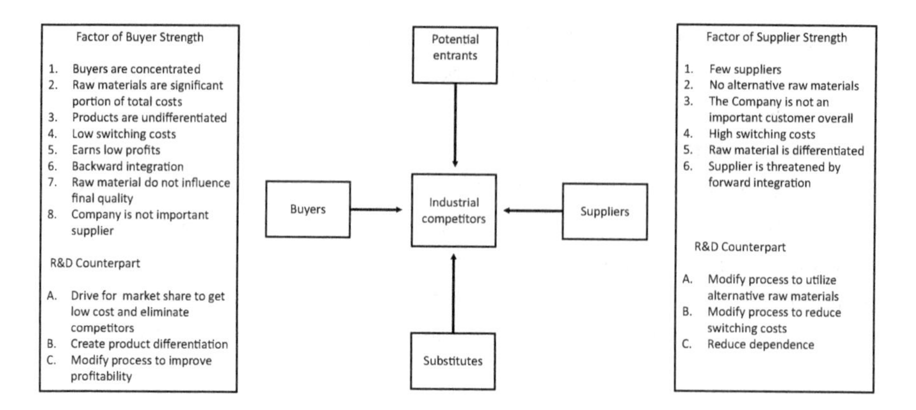

Figure 4.8. Project Generation from Threats from "On-going Flow of Products Dimension."

Source: Quintella *et al.* [1997, p. 124].

of products of the same type. At times, the industry boundary itself may be modified due to technology.

Also at this meso-level of analysis, Quintella *et al.* [1997] suggest a systematic approach for using business strategies as a source for research and development ideas (Fig. 4.8)

When referring to the technological strategy of the company, it is necessary to continue making reference to the work of Michael Porter. Porter referred to the technological strategy as everything related to the development and use of technology in the company, taking into account not only the research and technological development but also the technologies that can be used in the value chain. Porter sees technology as a key part of the company's strategy, but not the only one.

According to Porter, the technological strategy of a company must refer to three elements:

- decide what technologies to develop;
- decide whether to seek leadership in those technologies;
- the role of licensing technology.

To answer the question of what technologies we must develop, it is therefore necessary to think in terms of competitive advantage. In this sense, the technologies that the company develops must be in line with the generic strategies that this pursues. All the effort that the company develops in research and technological development should be oriented to the achievement of a strategy of differentiation, leadership in costs, or focus.

Product policies are usually identified with differentiation strategies, while process policies are more closely linked to cost strategies. However, this is not the case, since both the technological change in product or process can be oriented indistinctly to achieve strategies of differentiation, leadership in costs, or focus. For example, a new product design can reinforce its differentiation, but it can also reduce its costs due to less use of materials or the removal of parts of it, or even adapt to the specific needs of a segment of the market. On the other hand, technological change in processes can lead to greater efficiency and therefore a reduction of costs, but it is also possible to implant new methods of production or delivery of the product that reinforces its differentiation in the market.

It is also necessary that the company surpasses the vision of technology oriented only to its products or processes, since there are many technologies that are relevant throughout the value chain of the company. In this sense, it is interesting to analyze the technologies related to information systems, distribution, transport, etc., which have been presented at the beginning of this chapter. The company, therefore, should focus its efforts on those technologies that may have a more relevant effect on the company's differentiation or cost leadership strategy and in which it can maintain that effort in a sustainable way.

According to Porter, the second question we must answer is whether or not we should seek technological leadership. According to Porter, technological leadership is identified with the fact that a company seeks to be the first to introduce technological changes that support its competitive strategy. Porter comments that often companies that do not follow a strategy of technological leadership are seen as technological followers, although this is not always the case. The strategy of technological followership should be a conscious strategy and not the result of an indifference toward technological change.

The decision to pursue a technology leadership strategy can be implemented either through differentiation or through cost leadership. Porter illustrates these options in Table 4.1 [Porter, 1985, p. 181].

Table 4.1. Technological Leadership and Competitive Advantage.

	Technological leadership	Technological followership
Cost advantage	Pioneer the lowest cost product design. Be the first firm down the learning curve. Create low-cost ways of performing value activities.	Lower the cost of the product or value activities by learning from the leader's experience. Avoid R&D costs through imitation.
Differentiation	Pioneer a unique product that increases buyer value. Innovate in other activities to increase buyer value.	Adapt the product or delivery system more closely to buyer needs by learning from the leader's experience.

Source: Porter [1985].

The decision to choose a technology leadership strategy will depend on the sustainability of the technological leadership, the advantages the company can gain as a technology leader, and the disadvantages it faces. The interaction of these three factors is what will determine the choice of strategy. The sustainability of technological leadership will be favored by the difficulty that competitors would find in trying to imitate it and by the pace of technological change that the company can deploy in the market.

Porter posits that the sustainability of technological change is a function of the source of the technology, the presence or absence of a sustainable cost of differentiation advantage, relative technological skills, and the rate of technological diffusion. Regarding the source of technological change, the sustainability of such a change appears to be more complicated when it occurs outside the industry in question, for example, through suppliers. External sources minimize the importance of research and development within the firm, since they facilitate equal access to external technological resources. Second, the existence of cost or differentiation advantages favors the sustainability of this strategy. This is the case for economies of scale and learning effects that give companies with high costs in R&D the possibility to pay-off the investments in technological change. Third, companies with high relative technological skills are more likely to sustain technological leadership. High technological skills favor a more efficient use of R&D resources, improving the output from a given spending level on technology. Finally, the diffusion rate of technology will also influence the sustainability of the strategy. The company will try to delay the rate of exposure of its technology in the market through patents, retention of staff, and in-house developments, in such a way as to limit imitation by competitors.

In reference to the advantages or disadvantages of first movers, Porter establishes the following, which are summarized in Table 4.2.

It is very important for companies pursuing a technological leadership strategy to determine the lead time they want to establish with their competition [Rieck and Dickson, 1993]. Companies following a proactive technology strategy focus their research and development on state-of-the-art innovation projects while combining the advantages of the first movers. Firms often use proactive technology strategies to achieve economies of scale, establish themselves as industry standards, or control distribution channels [Lieberman and Montgomery, 1988]. These authors state that there are two basic mechanisms to take advantage of from the first mover strategy: advantages from learning or experience and advantages from success in patenting, both in product or process technologies. Thus, they state, "In the standard learning-curve model, unit production costs fall with cumulative output. This generates a sustainable cost advantage for the early entrant if learning can be kept proprietary and the firm can maintain leadership in market share." They add, "When technological advantage is largely a function of R&D expenditures, pioneers can gain advantage if [the] technology can be patented or maintained as trade secrets" [Lieberman and Montgomery, 1988, pp. 42–43].

Finally, technology licensing is the last point that reflects the chosen technological strategy. Being aware that a technology is a source of competitive advantage, licensing is an option that must be only taken into account under special conditions, since it may hamper competitive strategy.

However, there are some situations in which licensing a technology could be a feasible option. First, a company could consider licensing when the firm is not able to exploit it due to a lack of resources or skills. Many times, if the firm does not license, it triggers competitors to develop their own technologies based on the pioneer's solution. Second, a company may license a technology to tap unavailable markets, gaining

Table 4.2. Advantages and Disadvantages of First Movers.

First mover's advantages	First mover's disadvantages
Reputation: A company that is a pioneer in the market acquires the reputation of being the first, which is difficult to imitate by the competition.	*Pioneering costs*: First movers face costs such as regulatory approval, buyer's education, developing infrastructures, etc.
Pre-empting a positioning: The first mover can choose the most attractive product or market positions.	*Demand uncertainty*: Bearing the risk of first demand, putting capacity in place first.
Switching costs: The pioneer has the opportunity to set up switching costs that can avoid buyers to change to another value proposition.	*Changes in buyer needs*: The pioneer is vulnerable to changes in buyers' needs, which can make the technology obsolete.
Channel selection: The innovator can choose better channels and negotiate in better conditions than followers who are less attractive to the market.	*Specificity of investments to early generations or factor costs*: Innovator faces specific investments in current technology, which might not be valid for future development.
Proprietary learning curve: The first mover can obtain early learning curve effects, including cost advantages.	*Technological discontinuities*: A major switch in technology can make the technology deployed by the pioneer obsolete.
Favorable access to scarce resources: The pioneer has early access to facilities, inputs, and scarce resources because it contracts them before the competition.	
Definition of standards: First mover's products could become the market's standards, forcing late movers to adopt them.	
Institutional barriers: Innovators can set barriers against imitation through patenting or governmental agreements.	
Early profits derived from the position in the market that allow the pioneer to harvest profits.	

Source: Porter [1985].

some revenue from it that would not otherwise be possible. Third, sometimes licensing is the way to accelerate the pace of adoption of a technology, which can lead to its acceptance as the new market standard. Fourth, when an industry or market is unattractive for the company, licensing is a way to reap some royalties from the market instead of developing a modest position in it. Fifth, licensing is a way of creating good competition that can in return play an important role in blocking new entrants, sharing development costs, or stimulating demand. Finally, a company may license a technology as a *quid pro quo* in return from another firm's technology.

Porter suggests a number of analytical steps to implement a technological strategy, to "turn science into a competitive weapon" [Porter, 1985, p. 198].

(1) Identify all the distinct technologies and subtechnologies in the value chain. The first step is just to make a list of relevant technologies, not only those crucial in the company's value chain but also all the technologies that are used by the company or its competitors in the industry. It is also worth describing all the technologies used by the providers or buyers of the company which could also be interrelated with those of the company.
(2) Identify potentially relevant technologies in other industries or under scientific development. Very often technological change comes from outside the boundaries of the industry, triggering discontinuous change and competitive disruption in the industry. Every step in the value chain must be examined, considering whether existing technologies outside the industry could be applied.
(3) Determine the likely path of change to key technologies. The company must foresee the more likely technological path of change of key technologies used in its value chain and that of its buyers and providers.
(4) Determine what technologies and potential technological changes are more relevant for a competitive advantage and the competitive structure. Not all technologies are key to determining the competitive structure of the industry. Therefore, the company must seek technologies that can create sustainable competitive advantages themselves, shift costs or uniqueness drivers in the firm, lead to first mover's

advantages, and improve the whole structure of the industry. The company must find out how these technologies can affect costs or differentiation in the industry.

(5) Assess a firm's relative capabilities in important technologies and the costs of making improvements. The firm must evaluate its capabilities in existing relevant technologies and its capacity to keep up with technological change in a specific realm.

(6) Select a technology strategy, encompassing all important technologies that reinforce the overall competitive strategy of the firm. By doing so, the company should first determine the list of R&D projects in which it is going to be involved, specifying in every case whether they support differentiation or cost leadership. Second, the company must choose between technological leadership and followership. Third, it is necessary to establish a technology licensing policy, and finally, the means to obtain technology externally.

(7) Reinforce business unit technologies at a corporate level. Once the company has determined the list of key enabling competitive technologies, it should also analyze the corporate-level strategy technologies. These technologies impact more than one business unit and are important for the whole corporation itself.

Chiesa and Mazini [1998, p. 115], citing the works of Chiesa *et al.* [1996], Hamel and Prahalad [1994], and Prahalad and Hamel [1990], defend that "a different approach is needed in arenas with high innovation rates, very dynamic areas in which technology makes product paradigms obsolete in a short time, and the introduction of new product/market combinations is the basis for survival" and add, "In a highly dynamic environment, firms search for continuity, which cannot be associated with the product. Firms need to find continuity and coherence in terms of the skills and knowledge used for product application." These authors refer to the cumulative character of technological development, where the firm is acquiring a series of resources and competences over time. Therefore, they defend a learning orientation for the development of technological competencies, and the sustainability of the competitive advantage depends on the ability to accumulate technological capabilities over time, in such a way that these are difficult to imitate.

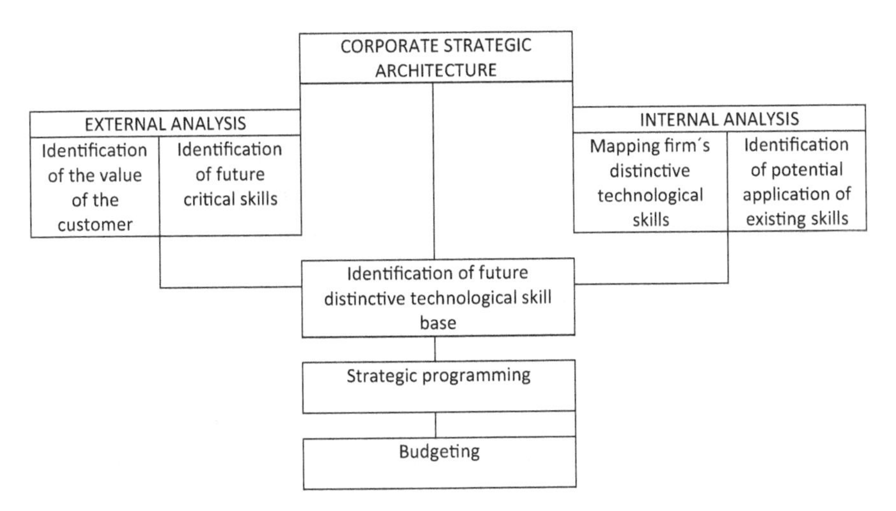

Figure 4.9. A Resource-based Framework for the Technology Strategy Formulation.
Source: Chiesa and Mazini [1998, p. 116].

These authors propose a methodology for the design of the technological strategy, which is summarized in Fig. 4.9 and is explained as follows:

- *External analysis*: The first step in the external analysis is to understand the value given to the consumer and its evolution over time. The value contributed by the product depends on the functionalities of the product, where the product is seen as a technological solution to satisfy the needs of the customer. Through the external analysis, we find out not only the current needs but the consumer behavior as well, to understand their future needs. The second step in the external analysis is to identify the company's competences and skills needed to fulfil customer's needs both now and in the future.
- *Internal analysis*: Being aware that in high competitive arenas, end product solutions change significantly over time, it is necessary to move from the analysis of output variables to state variables, such as knowledge and skills. Thus, this analysis includes the skill base mapping of the firm, skill benchmarking against competitors, and the identification of critical skills (high value for the customer, applicability, and appropriability).

- *Identification of the future distinctive technological skill base and plan of action*: This step focuses on the comparison between external and internal analysis to select the actions characterizing the firm's strategy. To accomplish this, it is necessary to make a distinction between applications and skills, and at the same time divide them into those which already exist within the firm and those new to the firm. In this way, we can build a matrix that defines the strategic actions undertaken by the company.

Consequently, a series of actions can be carried out in order to fulfil the company's technological strategy. These actions entail competence deepening, competence fertilizing, competence complementing, competence destroying, and competence refreshing (Fig. 4.10).

- *Competence deepening*: This means delving into the current skills and abilities needed to support the current strategy. This strategy will be feasible if the current skills are applicable and highly appropriable.

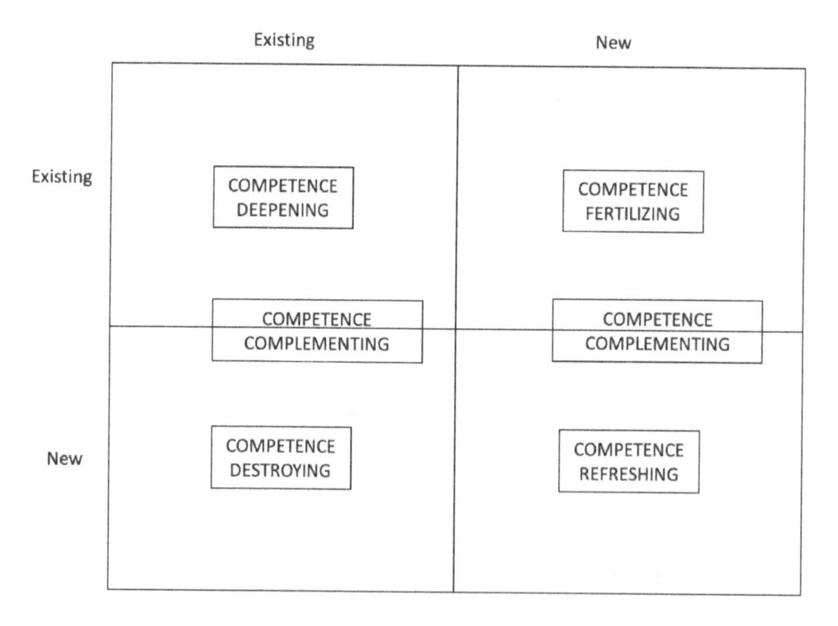

Figure 4.10. Technology Strategy Actions.

Source: Chiesa and Mazini [1998, p. 118].

Applying this strategy means relying on the current skills and the existing competition gap. A good way to carry out this strategy is to develop in-house R&D to strengthen the base of the current skills and competences.

- *Competence fertilizing*: This type of strategy implies being able to take advantage of the current skills base to support new applications. Synergies can occur with current applications, since they use the same competency base. Conversely, there may be problems on the marketing side, due to the company's lack of experience in these new applications. This fertilization strategy can be achieved through R&D programs or through agreements and alliances with third parties.
- *Competence complementing*: This strategy involves the acquisition of new skills and abilities to be combined with the current ones in such a way that they can take advantage of new markets or applications. This strategy involves progressively changing the existing competency base to a new one. New competencies can be developed internally or built through agreements or alliances.
- *Competence refreshing*: This strategy aims to build a new competence base to harness new applications in new markets. This kind of strategy can be highly risky, as it implies building up a new competence and skills base and new applications. A way to acquire this new base of competences may be the acquisition of a company in the area, internal ventures, or investments supported by internal R&D in a later phase.
- *Competence destroying*: This kind of strategy does not differ significantly from the refreshing strategy. It means that a new set of competences have come up to drive an existing set of applications. To keep up the pace in the market, the company needs to build up a new competence base and divest the former competence base.

Markides [1997], in his article "Strategic Innovation," proclaims the "virtues of breaking rules." The author refers to the work of Abell [1980] who distinguishes three questions to define a business: What products or services are you offering to your clients? Who is going to be your customer? And finally, how should the company offer these products or services in a cost-efficient way? This third question is meant to describe the technology used by the company. For Markides, strategic innovation is the way in which a company proactively and systematically thinks about and develops a new game plan. Consequently, it changes the rules of the game in the market. To do this, the author argues, the company can redefine the business itself, redefine the who (who is the customer, trying to identify new segments to serve), redefine the what (thinking about what products or services we are offering these customers), redefine the how (leveraging existing competences to build new products or a better way of doing business), and, finally, the author proposes to start this process from different angles.

Kumar *et al.* [2000] distinguish between market-driven and market-driving companies. The authors argue that, while companies are constantly required to be more market driven, the results from their research on 25 pioneering companies show that their success is based on the introduction of radical innovations, which could be defined as market-driving business. The authors state that market-driven businesses are more likely to achieve incremental innovations, though they occasionally produce radical innovations that change the rules of the market. Conversely, market-driving companies, which are normally new entrants in the market, gain more sustainable competitive advantages based on the uniqueness of their offer.

4.5 TOOLS FOR INNOVATION AND TECHNOLOGY MANAGEMENT

The management of innovation and technology has become a discipline with a growing body of knowledge. Throughout the literature, we can see how the concept has evolved from the "management of research and development" to the concept of "innovation management," "technology management," and finally "strategic management of technology" [Drejer, 1996]. "The evolution of TM is observed to take place from a stable and predictable situation within an R&D department to a discontinuous and unpredictable situation taking place at the strategic level" [Cetindamar *et al.*, 2009, p. 238]. This concept has received widespread attention from both business practitioners and scholars [Drejer, 1997] trying to define what technology management is, what tools and techniques are available, and what their functions are in supporting technology management within organizations.

Liao [2005, p. 381], based on *Task Force on Management of Technology* [1987], defined technology management as a "process, which includes planning, directing, control and coordination of the development and implementation of technological capabilities to shape and accomplish the strategic and operational objectives of an organization." Wang [1993] argues that technology management encompasses activities such as planning for the development of technology capabilities, the identification of key technology and its related fields for development, determining whether "to buy" or "to make"; i.e., whether importation or self-development should be pursued and the establishment of institutional mechanisms for directing and coordinating the development of technology capabilities, and the design of policy measures for controls.

As we have seen in the earlier section, we consider innovation and technology part of the strategic process of the company and key to achieving and sustaining a competitive advantage. In this way, the implementation of methods and tools for the management of innovation and technology in the company from a strategic approach is also key to the success of the company. In line with this strategic orientation of technology management, Linn *et al.* [2000] argue that the management of technology should not only refer to a certain technological field but also take into account the implementation of strategies according to available resources, current technologies, future markets, and the socioeconomic environment.

However, these models are varied, depending on the type of company, sector, or characteristics of the organization. In this section, we will take a tour of the main techniques of innovation and technology management, although in later chapters we delve into certain aspects such as the development of new products and processes, leadership and company culture, learning, and knowledge management.

Phaal *et al.* [2006, p. 336] state that, "Industrial and academic interest in how to more effectively manage technology is growing as the complexity, cost and rate of technological innovation increase, at a time of increasing organizational and industrial change on a global scale." Consequently, the authors proposed a technology management framework as shown in Fig. 4.11 [Phaal *et al.*, 2004].

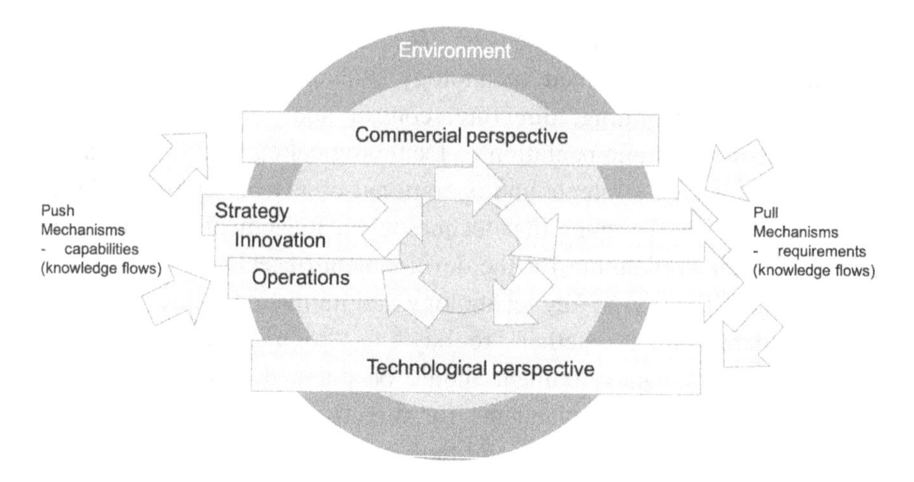

Figure 4.11. Technology Management Framework.
Source: Phaal *et al.* [2004, p. 7].

At the center of the framework we can see the technological base of the organization, which represents the technological knowledge, skills, and capacities that support the development and delivery of products and services. In this way, there are five management processes that operate in the technological base, such as the identification of technologies, and the selection, acquisition, exploitation, and protection of these technologies. The ISAEP processes are related to the organization's strategy, innovation, and operations. The authors defend that, "The aim of effective technology management is to ensure that technological issues are incorporated appropriately into these processes, to form a technology management system that is coherent and integrated across and beyond specific business processes and activities" [Phaal *et al.*, 2004, p. 8]. The framework depicts a balance between commercial and technological requirements, which must be fed by the knowledge flows that enter the system through pull and push processes. In summary, the technology management framework encompasses the following key elements: The technology base of the firm, the technology management process that operates on the technology base to support innovation, the core business processes of strategy (innovation and operations), the mechanism by which commercial and technological perspectives of the firm are brought together (market pull and technology push), internal and external factors (business culture, purpose, organizational structure, and economic environment), and time as a key dimension in technology management.

These authors made a classification of technology management tools distinguishing between matrices, grids, tables, and scored profiles. At the same time, they categorize the different tools into 11 different groups as shown in Table 4.3.

Another categorization of innovation and technology management tools has been developed by Hidalgo and Albors [2008]. The authors, based on the works of Thorn [1990], Cordero [1991], Ram [1996], European Commission [1996], European Commission [2005], Libutti [2000], Scozzi *et al.* [2005], and Phaal *et al.* [2006], developed a list of innovation management techniques and typologies, and the methodologies and tools associated with them. Thus, the authors distinguish between the IMT typologies and the associated methodologies as shown in Table 4.4.

It is also interesting to cite the work of Liao [2005] who posits a technology management framework and its applications. Liao proposed the following fields of application:

- Computer integrated manufacturing
- Construction project management
- Business process re-engineering
- Project appraisal
- Product design
- Disaster management
- Technology assessment
- Process design
- Engineering design
- Knowledge management

Levin and Barnard [2008] proposed a framework of technology management routines. These authors based their proposal on the work of Pavitt [2002]. Pavitt suggested dividing innovation into three different overlapping dimensions: the production of scientific and technological knowledge, the transformation of knowledge into working artefacts, and

Table 4.3. Technology Management Tools Catalogue.

Groups	Content
Technology management in the business, including portfolio, strategy, acquisition, and R&D management	Portfolio methods for strategy selection Technology strategy Grids for linking technology to the business Technology and management Technology acquisition and sourcing R&D management
Innovation management in the business	Innovation and the management of innovation
Knowledge management in the business, including learning and the management of IT systems	Knowledge and the management of knowledge Learning and organizational development Management processes and knowledge flows Information systems and technology
New product, service, and process development, which includes the management of the development of new products, services, and processes.	New product development Product creation and design Production processes/manufacturing
Business strategy including strategy development and deployment in the business	Portfolio methods for strategy and selection, Structure and process, Competitive sector/industry assessment.
Business management, which includes general management of the business, including topics such as leadership, e-commerce, sustainability, and globalization	Leadership E-commerce and the Internet Sustainability and environmental management International business Management and business models
Marketing and customers, including customer, brand, and product management	Segmentation of markets and customers Brand management Product management
Management of behavior, culture, and human resources	Organizational culture Human behavior Human resource management
Organizational design and collaboration	Organizational structure and design Alliances and other forms of collaboration
Change management, including planning and projects	The nature of change Human response to change Business and change management Planning and project management
General problem solving and decision making	Nature, structure, and classification of problems Approaches to problem solving

Source: Phaal *et al*. [2006].

Table 4.4. IMT Typologies and the Associated Methodologies.

IMT typologies	Methodologies and tools
Knowledge management tools	Knowledge audits/Knowledge mapping/Document management/IPR management
Market intelligence techniques	Technology watch/Technology search/Patents analysis/Business intelligence/CRM/ Geo-marketing
Cooperative and networking tools	Groupware/Team-building/Supply chain management/Industrial clustering
Human resources management techniques	Tele-working/Corporate intranets/Online recruitment/ e-Learning/Competence management
Interface management approaches	R&D — Marketing interface management/ Concurrent engineering
Creativity development techniques	Brainstorming/Lateral thinking/TRIZ/Scamper method /Mind mapping
Process improvement techniques	Benchmarking/Workflow/Business process re-engineering/Just-in-time
Innovation project management techniques	Project management/Project appraisal/Project portfolio management
Design and product development management tools	CAD systems/Rapid prototyping/Usability approaches/Quality function deployment/Value analysis
Business creation tools	Business simulation/Business plan/Spin-off from research to market

Source: Hidalgo and Albors [2008].

matching artefacts with user's requirements either internal, such as process innovations), or external, such as product innovation. Levin and Barnard [2008] in their proposal of a framework for technology management routines, are aware of the considerations from Van de Ven and Angle [1989] and Van de Ven [1999], considering that technology management routines do not operate in a linear sequence. Thus, we can see the different technology management routines suggested by these authors (Table 4.5).

Cetindamar *et al.* [2009] integrate the theory of dynamic capabilities into the technology management framework developed by Phaal *et al.*

Table 4.5. Framework of Technology Management Routines.

Producing scientific and technological knowledge	Transforming knowledge into working artefacts	Matching artifacts with user's requirements
Ideation (creative process to develop new products/ processes).	Technology road mapping (develop a plan for what technologies would be needed to support a given product/process in the future).	Business unit environmental monitoring (scan and analyze competitors, suppliers, customers, technologies, regulators, etc.).
R&D environmental monitoring (scan and analyze the external environment, especially technology).	Product line planning (develop a plan for future direction of product line/ platform).	Corporate environmental monitoring (scan and analyze competitors, suppliers, customers, technologies, regulators, etc.).
R&D technology strategy (plan progression of technology to be developed by R&D).	Product portfolio management (evaluate portfolio of products to achieve balance along dimensions).	Business unit business strategy (develop the business unit's plan and budget).
R&D portfolio management (evaluate portfolio of R&D projects to achieve the desired balance along different dimensions).	Feasibility (investigate the market and technical feasibility of an idea).	Corporate business strategy (develop the company's overall plan and budget).
Intellectual property management (manage patents, copyright, trademarks, standards).	Project execution (planning, designing, staffing, and managing the actual work of a project).	Technology needs assessment (determine what technologies current and future customers want).
Post-project audit (discuss and disseminate lessons learned).	Technology transfer (shift ownership of artifacts and accompanying knowledge).	Business unit technology strategy (determine the role of various technologies in a business unit).

(*Continued*)

Table 4.5. (*Continued*)

	Technology adaptation (absorb and adapt technical artifacts and accompanying knowledge).	Corporate technology strategy (determine the role of R&D and technology in company).
	Post-project support (provide support to adopters of technology).	Initial program/project selection (determine if a program/project should be funded).
		R&D funding (determine how to fund R&D efforts).
		New business unit development (determine when a new set of products/technologies/markets warrant the formation of a new business unit).
Providing organizational support.		
Performance management (measure and manage performance).		
Personnel management (hire and develop skilled personnel).		
Technology alliance management (identify, develop, and manage strategic partnerships and consortia).		

Source: Levin and Barnard [2008].

[2004, p. 242], proposing a technology management model based on six main activities as follows:

- Identification of technologies that are or may be important for the business. The activities included are searching, auditing, data collection, and intelligence processes.
- Selection of technologies linking strategic vision with assessment of technological capabilities. Selection includes setting strategic objectives and alignment of technology and business strategies.
- Acquisition of selected technologies, including the choice among buy, collaboration, or own elaboration.

- Exploitation of technologies. First, whether technology comes from internal development or external acquisition, it is necessary to assimilate the technology. Thereafter, exploitation refers to commercialization of the new technology embedded in new products or implemented in new processes in the firm.
- Protection of knowledge, through patenting or staff retention.
- Learning from the development and exploitation of technologies, linking it to the process of knowledge management.

4.6 TECHNOLOGY SOURCING AND STRATEGY OF THE FIRM

We have referred to the role of technology in the company's strategy. However, it is interesting that we analyze the sources of technology, that is, whether they come from internal developments or are acquired from the outside through purchase or agreements with third parties. The determinants of the choice between the internal development of technology and through external sources have been investigated from the perspective of transaction costs [Mowery *et al.*, 1998; Robertson and Gatignon, 1998] and from knowledge-based theories [Grandori and Kogut, 2002].

The internal generation of technology depends on the firm's R&D capability. In many organizations, R&D is considered a general expense of the company aimed at enhancing efforts to maintain the necessary knowledge in areas relevant to the organization. Usually, the company's R&D effort is measured in terms of the percentage of sales that it dedicates to the R&D budget or where the company has R&D resources and facilities. Harmsen *et al.* [2000] argue that the internal development of technology is a risky option, since it does not guarantee a positive result by generating new knowledge or an application in the market [Cohen and Levinthal, 1990]. However, internal sourcing can generate proprietary technology that may constitute the base for a competitive advantage of the firm, developing a strong technological base of the firm that is hardly imitable [Howells, 1997].

Another option for the firm is going outside the company boundaries to get the technological assets and capabilities needed for their strategy, this being one of the main drivers to develop collaborative agreements to obtain resources required to build competitive advantages [Bruce *et al.*, 1995]. On the one hand and, according to the transactional cost theory, there are some barriers hampering the process of external sourcing, such as partner selection and coordination costs [Pisano, 1990; Steensma and Corley, 2001]. Moreover, taking into account that the technology available in the market is also accessible to competitors, it may not constitute the base of a competitive advantage [Montoya *et al.*, 2007]. On the other hand, the benefits of sharing costs and risks of R&D [Huang *et al.*, 2009] and risk of knowledge leakages and imitation [Mowery *et al.*, 1998] must be considered advantages of external sourcing.

The motivation to acquire technology outside the firm comes from the possibility of developing technological capabilities, developing strategic options, gaining efficiency, and responding to the economic environment [Ford *et al.*, 2010]. The pursuit of technology outside the company can be carried out through licensing, strategic technological agreements, and mergers and acquisitions, or through the search for personnel with the necessary technological know-how.

It seems that to expand the possibilities of innovation within companies, they are willing to combine the internal competencies of the company with complementary knowledge from outside the firm [Arora and Gambardella, 1990].

However, the decision to pursue externally sourced technology is not only a question of identifying the right technology but also understanding how this technology will be integrated into the company's strategy [Nesta and Dibiaggio, 2003]. Accordingly, Jiménez-Barrionuevo *et al.* [2011] argue that technology acquisition is related not only to the identification and assimilation of the technology but also to its application to practical ends.

The acquisition of technology follows different steps, such as the definition of technology requirements, the identification of available technologies, technology evaluation and selection of the source, negotiation, implementation and value-addition, and relationship management [Ortiz-Gallardo, 2013].

The acquisition of external technology aims to establish barriers for competitors to defend a strategic position of the firm, usually through the development of new competences by means of technology inputs. By way of explanation, it is worth citing the resource-based view (RBV), which argues that firms are a collection of tangible and intangible resources [Das and Teng, 2000]. In particular, when these resources are valuable, rare, inimitable, and non-substitutable, they lead to unique capabilities [Barney, 1991, cited in Ortiz-Gallardo, 2013].

Barney [1991] argues that there are three kinds of resources: physical resources, which include tangible assets of the firm; human resources, which comprise skills, experience, and training of company's staff; and finally, organizational resources, which encompass corporate culture, the firm's relationships, procedures, and organizational structure. Das and Teng [2000] suggest that some resources can be sorted in two categories: property-based resources and knowledge-based resources. Moreover, these authors identify valuable resources as scarce and lacking in direct substitutes, normally mixed with other resources, and embedded in organizational structures.

From this point of view, technology can be described as a valuable resource for the company, which can be used to build a competitive advantage.

REFERENCES

Abell, D. F. [1980]. *Defining the Business: The Starting Point of Strategic Planning* (Prentice-Hall, Englewood Cliffs, NJ).

Abernathy, W. J., and Utterback, J. M. [1978]. Patterns of industrial innovation, *Technology Review*, 80(7), pp. 40–47.

Abernethy, M. A., and Brownell, P. [1999]. The role of budgets in organizations facing strategic change: An exploratory study, *Accounting, Organizations and Society*, 24(3), pp. 189–204.

Anderson, P., and Tushman, M. L. [1990]. Technological discontinuities and dominant designs: A cyclical model of technological change, *Administrative Science Quarterly*, pp. 604–633.

Andrews, K. R. [1971]. *The Concept of Corporate Strategy* (Dow Jones, Irwin, NY).

Arora, A., and Gambardella, A. [1990]. Complementarity and external linkages: The strategies of the large firms in biotechnology, *The Journal of Industrial Economics*, pp. 361–379.

Barney, J. [1991]. Firm resources and sustained competitive advantage, *Journal of Management*, 17(1), pp. 99–120.

Becker, G. S. [1964]. *Human Capital* (Columbia, New York).

Bone, S., and Saxon, T. [2000]. Developing effective technology strategies, *Research-Technology Management*, 43(4), pp. 50–58.

Bruce, M., Leverick, F., and Littler, D. [1995]. Complexities of collaborative product development, *Technovation*, 15(9), pp. 535–552.

Cetindamar, D., Phaal, R., and Probert, D. [2009]. Understanding technology management as a dynamic capability: A framework for technology management activities, *Technovation*, 29(4), pp. 237–246.

Chandler, J. A. D. [1962]. *Strategy and Structure: Chapters in the History of the American Industrial Enterprise* (Beard Books, Washington, D.C.).

Chiesa, V., Coughlan, P., and Voss, C. A. [1996]. Development of a technical innovation audit, *Journal of Product Innovation Management*, 13(2), pp. 105–136.

Chiesa, V., and Mazini, R. [1998]. Towards a framework for dynamic technology strategy, *Technology Analysis & Strategic Management*, 10(1), pp. 111–129.

Christensen, C. M. [1992]. Exploring the limits of the technology S-curve. Part I: Component technologies, *Production and Operations Management*, 1(4), pp. 334–357.

Christensen, C. M. [1993]. The rigid disk drive industry: A history of commercial and technological turbulence, *Business History Review*, 67(04), pp. 531–588.

Christensen, C. M. [1997]. *The Innovator's Dilemma* (Harvard Business School Press, Boston).

Christensen, C. M., Suárez, F. F., and Utterback, J. M. [1998]. Strategies for survival in fast-changing industries, *Management Science*, 44(12, Part 2), pp. 207–220.

Clarke, K., Ford, D., and Saren, M. [1989]. Company technology strategy, R*&D Management*, 19(3), pp. 215–229.

Cohen, W. M., and Levinthal, D. A. [1990]. Absorptive capacity: A new perspective on learning and innovation, *Administrative Science Quarterly*, 35(1), pp. 128–152.

Coombs, R., and Richards, A. [1990]. *The Integration of R&D Strategy and Business Strategy* (CRONTEC, Manchester School of Management, Manchester).

Cordero, R. [1991]. Managing for speed to avoid product obsolescence: A survey of techniques, *Journal of Product Innovation Management*, 8(4), pp. 283–294.

Daft, R. L. [2010]. *Organization Theory and Design*, South-Western Cengage Learning.

Das, T. K., and Teng, B.-S. [2000]. A resource-based theory of strategic alliances, *Journal of Management*, 26(1), pp. 31–61.

D'aveni, R. A. [1994]. *Hypercompetition,* 1st edn. (Free Press, New York).

Davenport, S., Campbell-Hunt, C., and Solomon, J. [2003]. The dynamics of technology strategy: An exploratory study, R&D *Management*, 33(5), pp. 481–499.

David, P. A. [1969]. *A Contribution to the Theory of Diffusion* (Research Center in Economic Growth, Stanford University).

De Meyer, A. [2008]. Technology strategy and China's technology capacity building, *Journal of Technology Management in China*, 3(2), pp. 137–153.

Dosi, G. [1982]. Technological paradigms and technological trajectories: A suggested interpretation of the determinants and directions of technical change, *Research Policy*, 11(3), pp. 147–162.

Drejer, A. [1996]. Frameworks for the management of technology: Towards a contingent approach, *Technology Analysis & Strategic Management*, 8(1), pp. 9–20.

Drejer, A. [1997]. The discipline of management of technology, based on considerations related to technology, *Technovation*, 17(5), pp. 253–265.

Drucker, P. F. [1994]. The theory of business, *Harvard Business Review*, pp. 95.

European Commission [1996]. *Innovation Management Tools: A Review of Selected Methodologies* (EIMS Publication 30. DG-XIII-D, Bruxelles).

European Commission [2005]. *Innovation Management Techniques* (IMT Toolbox). (Cordis, Bruxelles).

Ford, D., and Thomas, R. [1997]. Technology strategy in networks, *International Journal of Technology Management*, 14(6–8), pp. 596–612.

Ford, S., Garnsey, E., and Probert, D. [2010]. Evolving corporate entrepreneurship strategy: Technology incubation at Philips, R&D *Management*, 40(1), pp. 81–90.

Foster, R. N. [1986]. *Innovation: The Attacker's Advantage* (Summit Books, New York).

Fusfeld, A. R. [1978]. How to put technology into corporate-planning, *Technology Review*, 80(6), pp. 51–55.

Gebauer, H., Worch, H., and Truffer, B. [2012]. Absorptive capacity, learning processes and combinative capabilities as determinants of strategic innovation, *European Management Journal*, 30(1), pp. 57–73.

Grandori, A., and Kogut, B. [2002]. Dialogue on organization and knowledge, *Organization Science*, 13(3), pp. 224–231.

Grant, R. M. [2002]. *Contemporary Strategy Analysis: Concepts, Techniques, Applications* (Blackwell Publishers).

Grindley, P. [1995]. *Standards, Strategy, and Policy: Cases and Stories*, 1st edn. (Oxford University Press, Oxford).

Hamel, G., and Prahalad, C. K. (1994). *Competing for the Future* (Harvard Business Press).

Harmsen, H., Grunert, K. G., and Declerck, F. [2000]. Why did we make that cheese? An empirically based framework for understanding what drives innovation activity, *R&D Management*, 30(2), pp. 151–166.

Hax, A. C., and Majluf, N. S. [1984]. *Strategic management: An integrative perspective* (Prentice Hall).

Hax, A. C., and No, M. [1993]. *Linking Technology and Business Strategies: A Methodological Approach and an Illustration*, Sarin, R. (eds.), *Perspectives in Operations Management* (Springer, US) pp. 133–155.

Hidalgo, A., and Albors, J. [2008]. Innovation management techniques and tools: A review from theory and practice, *R&D Management*, 38(2), pp. 113–127.

Hofer, C. W., and Schendel, D. [1978]. *Strategy Formulation: Analytical Concepts* (West Publishing, St. Paul: South-Western).

Howells, J. [1997]. Management and the hybridization of expertise: EFTPOS in retrospect, *Journal of Information Technology*, 12(1), pp. 83–95.

Huang, Y.-A., Chung, H.-J., and Lin, C. [2009]. R&D sourcing strategies: Determinants and consequences, *Technovation*, 29(3), pp. 155–169.

Jensen, R. [1982]. Adoption and diffusion of an innovation of uncertain profitability, *Journal of Economic Theory*, 27(1), pp. 182–193.

Jiménez-Barrionuevo, M. M., García-Morales, V. J., and Molina, L. M. [2011]. Validation of an instrument to measure absorptive capacity, *Technovation*, 31(5), pp. 190–202.

Katz, E., Levin, M. L., and Hamilton, H. [1963]. Traditions of research on the diffusion of innovation, *American Sociological Review*, 28(2), pp. 237.

Kumar, N., Scheer, L., and Kotler, P. [2000]. From market driven to market driving, *European Management Journal*, 18(2), pp. 129–142.

Levin, D. Z., and Barnard, H. [2008]. Technology management routines that matter to technology managers, *International Journal of Technology Management*, 41(1–2), pp. 22–37.

Liao, S. [2005]. Technology management methodologies and applications: A literature review from 1995 to 2003, *Technovation*, 25(4), 381–393.

Libutti, L. [2000]. Building competitive skills in small and medium-sized enterprises through innovation management techniques: Overview of an Italian experience, *Journal of Information Science*, 26(6), pp. 413–419.

Lieberman, M. B., and Montgomery, D. B. [1988]. First-mover advantages, *Strategic Management Journal*, 9(S1), pp. 41–58.

Linn, R. J., Zhang, W., and Li, Z. [2000]. An intelligent management system for technology management, *Computers & Industrial Engineering*, 38(3), pp. 397–412.

Mansfield, E. [1961]. Technical change and the rate of imitation, *Econometrica: Journal of the Econometric Society*, pp. 741–766.

Mansfield, E. [1968]. *The Economics of Technological Change* (Norton, New York).

Markides, C. [1997]. Strategic Innovation, *Sloan Management Review*, 38(3).

Miles, R. E., Snow, C. C., Meyer, A. D., and Coleman, H. J. [1978]. Organizational strategy, structure, and process, *Academy of Management Review*, 3(3), pp. 546–562.

Montoya, P. V., Zarate, R. S., and Martín, L. Á. G. [2007]. Does the technological sourcing decision matter? Evidence from Spanish panel data, *R&D Management*, 37(2), pp. 161–172.

Mowery, D. C., Oxley, J. E., and Silverman, B. S. [1998]. Technological overlap and interfirm cooperation: Implications for the resource-based view of the firm, *Research Policy*, 27(5), pp. 507–523.

Nesta, L., and Dibiaggio, L. [2003]. Technology strategy and knowledge dynamics: The case of biotech, *Industry and Innovation*, 10(3), pp. 331–349.

Ortiz-Gallardo, V. G. [2013]. *Technology acquisition: Sourcing technology from industry partners* (University of Cambridge, Cambridge).

Oster, S. M. [1999]. *Modern Competitive Analysis* (Oxford University Press).

Pavitt, K. [2002]. Innovating routines in the business firm: What corporate tasks should they be accomplishing? *Industrial and Corporate Change*, 11(1), pp. 117–133.

Penrose, E. [2013]. *The Theory of the Growth of the Firm* (Martino Fine Books, Mansfield Centre, CT).

Phaal, R., Farrukh, C. J. P., and Probert, D. R. [2004]. A framework for supporting the management of technological knowledge, *International Journal of Technology Management*, 27, pp. 1–15.

Phaal, R., Farrukh, C. J., and Probert, D. R. [2006]. Technology management tools: Concept, development and application, *Technovation*, 26(3), pp. 336–344.

Pisano, G. P. [1990]. The R&D boundaries of the firm: An empirical analysis, *Administrative Science Quarterly*, pp. 153–176.

Porter, M. E. [1980]. *Competitive Strategy: Techniques for Analyzing Industries and Competitors* (Free Press, New York).

Porter, M. E. [1985]. *Competitive Advantage: Creating and Sustaining Superior Performance* (Free Press, New York).

Prahalad, C. K., and Hamel, G. [1990]. *Core Competence of the Corporation* (HBR Bestseller).

Quintella, R. H., Dias, C. C., and Vasconcelos, B. [1997]. Technology strategy formulation: AIDS, methodology and framework of analysis, *Organizações & Sociedade*, 4(10), pp. 117–132.

Ram, S. [1996]. Validation of expert systems for innovation management: Issues, methodology, and empirical assessment, *Journal of Product Innovation Management*, 13(1), pp. 53–68.

Reinganum, J. F. [1981]. On the diffusion of new technology: A game theoretic approach, *The Review of Economic Studies*, 48(3), pp. 395–405.

Rieck, R. M., and Dickson, K. E. [1993]. A model of technology strategy: Practitioners' forum, *Technology Analysis & Strategic Management*, 5(4), pp. 397–412.

Robertson, T. S., and Gatignon, H. [1998]. Technology development mode: A transaction cost conceptualization, *Strategic Management Journal*, pp. 515–531.

Rogers, E. M. [1962] *Diffusion of Innovations* (Free Press of Glencoe).

Rothaermel, F. [2008]. Competitive advantage in technology intensive industries, *Technological Innovation: Generating Economic Results*, Libecap, G. and Thursby, M. (eds.), Advances in Study of Entrepreneurship, Innovation and Economic Growth, Vol. 18 [Emerald Group Publishing Limited, Bingley], pp. 201–225.

Roussel, P. A., Saad, K. N., and Erickson, T. J. [1991]. *Third Generation R & D: Managing the Link to Corporate Strategy*. 1st edn. (Harvard Business Review Press, Boston).

Schlegelmilch, B. B., Diamantopoulos, A., and Kreuz, P. [2003]. Strategic innovation: The construct, its drivers and its strategic outcome, *Journal of Strategic Marketing*, 11(2), pp. 117–132.

Scozzi, B., Garavelli, C., and Crowston, K. [2005]. Methods for modeling and supporting innovation processes in SMEs, *European Journal of Innovation Management*, 8(1), pp. 120–137.

Shortell, S. M., and Zajac, E. J. [1990]. Perceptual and archival measures of Miles and Snow's strategic types: A comprehensive assessment of reliability and validity, *Academy of Management Journal*, 33(4), pp. 817–832.

Steensma, H. K., and Corley, K. G. [2001]. Organizational context as a moderator of theories on firm boundaries for technology sourcing, *Academy of Management Journal*, 44(2), pp. 271–291.

Szakonyi, R. [1990]. Coordinating R&D and business planning, *Technology Analysis & Strategic Management*, 2(4), pp. 391–412.

Task Force on Management of Technology [1987]. *Management of Technology: The hidden competitive advantage* (National Academy Press, Washington).

Thorn, N. [1990]. Innovation management in SMEs, *Management International Review*, 30(2), pp. 181–192.

Tushman, M. L., and Rosenkopf, L. [1992]. Organizational determinants of technological-change-toward a sociology of technological evolution, *Research in Organizational Behavior*, 14, pp. 311–347.

Twiss, B. C. [1992]. *Managing Technological Innovation* (Pitman, London).

Utterback, J. M. [1996]. *Mastering the Dynamics of Innovation*, 2nd edn. (Harvard Business Review Press, Boston).

Van de Ven, A. H. [1999]. *The Innovation Journey* (Oxford University Press, Oxford).

Van de Ven, A. H., and Angle, H. L. [1989]. *Suggestions for Managing the Innovation Journey* (Strategic Management Research Center, University of Minnesota).

Wade, J. [1995]. Dynamics of organizational communities and technological bandwagons: An empirical investigation of community evolution in the microprocessor market, *Strategic Management Journal*, 16(S1), pp. 111–133.

Wang, H. [1993]. Technology management in a dual world, *International Journal of Technology Management*, 8(1–2), pp. 108–120.

Weil, E. D., and Cangemi, R. R. [1983]. Linking long-range research to strategic planning, *Research Management*, 26(3), pp. 32–39.

Williamson, O. E. [1983]. *Markets and Hierarchies: Analysis and Antitrust Implications* (Macmillan, USA).

CHAPTER 5

ORGANIZATION, LEADERSHIP, AND CULTURE

5.1 THE ORGANIZATIONAL STRUCTURE FOR INNOVATION

There is a widespread belief that organizational structures have a direct impact on the innovative capacity of organizations. Organizational structures vary depending on the strategy of the firm and how the company faces the environment. Accordingly, to survive successfully over a long period of time facing environmental and technological changes, companies are required to adapt their structural alignments [Schumpeter, 1939; Tushman and O'Reilly, 2002].

The classical theories of organization — the seminal work of Taylor [1911] and his renowned theory of scientific administration, Fayol's administrative management theory, or the bureaucratic models [Weber, 1947] — have always tried to find the best ways to shape the company in order to fulfil the business's requirements. All the former models and theories mainly focused on seeking efficiency in the workplace through the division of activities, the assignment of roles, and a hierarchically based authority. These mechanistic models, based on the seeking of productivity, were viewed as more suitable for static and stable environments. These classical and bureaucratic models did not consider change or innovation as part of the principles to build up their theories, but they were focused on the internal organizational features of the firm and aimed at discovering the best way to organize procedures to fulfil the corporation's objectives.

Conversely, the contingency theory faces the organizational challenge in relation to different contingency variables, such as technology and environmental demands, which affect organizational structures and innovation management. Hence, this line of thought argued that organic organizational structures are better suited for dynamic and changing environments, adapting more rapidly to conditions of quick change and innovation [Burns and Stalker, 1961]. These organic forms are characterized by a lack of task definition, informal coordination mechanisms, and less formalization and specialization. The authors argued that differentiation (e.g., heterogeneity of occupational profiles), professionalism, decentralization, environmental uncertainty, and inter-organizational interdependence would be positively related to innovation. Conversely, firms that operate in stable environments develop mechanistic management systems characterized by hierarchical relations, well-defined roles and responsibilities, high formalization, stratification, and large size, which are all negatively related to innovation. This statement has been validated in the research of Aiken *et al.* [1980] and Covin and Slevin [1989]. At the same time, the hypothesis that structural forms depend on the strategy of the firm and the kind of environment that the company is involved in has been validated by an array of authors [Aldrich, 1999; Sine *et al.*, 2006; Tushman and O'Reilly, 2002].

Another relevant contribution came from Mintzberg [1979], who, in his work *The Structuring of Organizations*, established a series of archetypes of organizational structures. Mintzberg distinguished five different structural forms, including simple structures, machine bureaucracy, professional bureaucracy, divisionalized forms, and adhocracy. A description of each structural form and its feasibility for innovation are shown in Table 5.1.

Mintzberg [1980, pp. 336–337] argues that the best structural form for innovation is adhocracy. He stated that, "Sophisticated innovation requires a fifth and very different structural configuration, one that is able to fuse experts drawn from different specialties into smoothly functioning project teams. Adhocracy is such a configuration." Adhocracy is defined by the author as a little-formalized, organic structure with a horizontal job specialization that groups professionals in market-based teams, performs the mutual adjustment coordination mechanism, and takes care of decentralization. Following the author, there are two types of adhocracy. On the one

Table 5.1. Different Structural Forms and Their Roles in Innovation.

Structural forms	Characterization and role for innovation
Simple structures	Simple structures are organic archetypes usually controlled by one person that easily adapt to changes in the environment. These structural forms are suitable for highly innovative and risky environments, like small start-ups in high-technology sectors.
Machine bureaucracy	Machine bureaucracy is characterized by high levels of specialization and standardization and is centrally controlled. These structural forms are quite appropriate for efficiency and control, but they are not adequate for a quick response to changing environments.
Professional bureaucracy	This kind of form allows high degrees of autonomy in professionals, with a concentration of power in authorized experts. On the one hand, individual experts can be highly innovative. On the other hand, the difficulties of coordination among individuals may hamper innovation within the organization.
Divisionalized forms	Divisionalized forms are decentralized structures that function autonomously, linked only by an administrative structure. These forms are able to develop abilities in specific markets. On the contrary, knowledge sharing between divisions is difficult, not being the best option for innovation.
Adhocracy	Adhocracy is the more suitable structural form for innovation. This organizational structure is designed to deal with unstable and complex environments, professional partnerships being a good example in this vein. These structural forms have a great ability to learn and they are very adaptive and innovative.

Source: Mintzberg [1979].

hand, an "operating adhocracy" where "the innovation is carried out directly on behalf of the clients, as in the case of consulting firms, advertising agencies, and film companies"; and on the other hand, an "administrative adhocracy" where the "project work serves the organization itself, as in the case of chemical firms and space agencies."

Therefore, it seems to be a two-fold approach where companies choose either an organic or a mechanistic model, depending on the type of environment they are involved in. However, authors such as Lawrence and Lorsch [1967] recognize that both models can coexist in some organizations, developing hybrid models that are able to face both evolutionary and breakthrough technological changes or efficiency versus flexibility. This type of organization performs with a kind of ambidexterity as pointed out in the work of O'Reilly and Tushman [2004, p. 24]. The term ambidexterity is defined as "the ability to simultaneously pursue both incremental and discontinuous innovation...from hosting multiple contradictory structures, processes, and cultures within the same firm." Some authors talk about a trade-off between efficiency and flexibility as a paradox for administration [Thompson, 1967], while others identify exploitation with company efficiency and exploration with discovery, autonomy, and innovation [March, 1991]. Throughout the literature, ambidexterity has been revealed to be positively linked with innovation [Yang and Atuahene-Gima, 2007; Tushman *et al.*, 2010; Sarkees and Hulland, 2009; Phene *et al.*, 2012].

Another couple of concepts that are relevant to company structuring came from Burgelman and Sayles [1988], who argue that there are two organizational design factors relevant to the innovation stream, which are the strategic importance of the innovation idea (i.e., up to which extent the innovative idea is aligned with the strategy of the firm) and operational relatedness (i.e., how closely related production requirements are to the innovative idea). Taking into account both variables, Burgelman and Sayles [1988] depicted a framework of nine organizational forms that are most suitable for each situation, as you can see in Fig. 5.1.

Here is a short description of each organizational form and its relatedness to innovation:

- *Direct integration*: This form is feasible when both the existing operational structure and strategic importance are strongly related with the innovation (i.e., modifications or extensions of existing products).
- *New product business department*: This structural form is recommended for innovative projects with a very important strategic relevance for the firm and medium operational relatedness. The company may be using some existing capabilities, but there is a difference with respect to existing products or services (i.e., new product development). In this case, a new product business department can harness the existing capabilities and experience, but at the same time it is able to

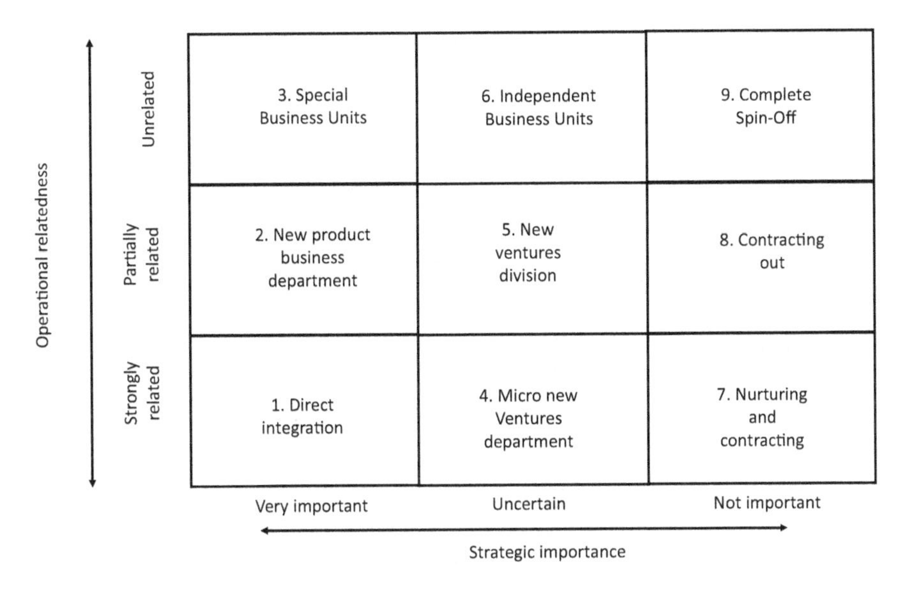

Figure 5.1. Organizational Structure and Innovation.
Source: Burgelman and Sayles [1988, p. 183].

organize new operational procedures partially independent from the general procedures of the firm.

- *Special business units*: When the company is involved in a strategically important innovation for the firm, but the operational relatedness is low (i.e., the firm is entering a new market), the company should set up a special business unit that is self-controlled. The company must develop new competences and capabilities and the way of doing is totally new for the firm.
- *Micro new ventures department*: There are some innovation projects that are not clearly related to the strategic mission of the company but very close to the operational procedures of the firm. In this case, a ventures department is an organizational solution to support this kind of innovation stream.
- *New ventures division*: It is not clear that the innovation stream must be strategically linked within the company. Neither is the operational relatedness clear. Therefore, the firm can create a division to manage the innovation independently from the core business in the long run.

- *Independent business unit*: In the case where strategic relevance is uncertain and there is a low operational relatedness, the most appropriate structure is an independent body.
- *Nurturing and contracting*: This kind of organizational structure is suitable for situations where operational relatedness is very high, but strategic importance is very low (i.e., a parent firm that decides to support a start-up idea, nurturing and subcontracting it).
- *Contracting out*: When the innovative stream is not strategically relevant and operational relatedness is partial (i.e., parent firm may take part in a company to which it is contracting out).
- *Complete spin-off*: Low strategic relevance and low operational relatedness. The best solution in this case is to create a spin-off to manage the innovation stream totally independent from the firm.

Furthermore, we have to make reference to one of the most important works in the literature studying the relationship between organizational structure and innovation, which is the research carried out by Teece [1996]. This author studied the characteristics of the innovation process itself in order to understand the organizational structure requirements of the firm in relation to the innovation process. Thus, the author recognized that the innovation process has some features, such as uncertainty, path dependency, cumulative nature, irreversibilities, technological interrelatedness, tacitness, and inappropriability. An explanation of each concept can be found in Table 5.2.

At the same time, the author considers a series of organizational and market determinants of the innovation performance of the firm as follows:

- M*onopoly power*: On the one hand, competence and rivalry are important for innovation, because high competition rates push firms to differentiate through the implementation of new products or processes. On the other hand, fragmented markets using non-proprietary technologies do not seem to be ideal for innovation, since homogeneity in means and resources hinders the capacity to innovate. Innovation clearly needs capital, which comes from cash flow or equity. Moreover, monopolistic situations generate the necessary cash flow to invest in innovation.

Table 5.2. Characteristics of the Innovation Process.

Characteristics of the innovation process	Explanation
Uncertainty	Related to the innovation concept itself. The author defines it as "a quest into the unknown."
Path dependency	Makes reference to the evolution of technology in certain path-dependent ways.
Cumulative nature of the innovation process	Technology development is cumulative along a defined path.
Irreversibility	New developments close the door to the competition based on older technologies.
Technological interrelatedness	The linkages between different organizational subsystems.
Tacitness	The nature of the knowledge developed by organizations and its difficulty to be transferred throughout the firm, being necessary to develop organizational routines as technological capabilities.
Inappropriability	The difficulties of new knowledge protection.

Source: Teece [1996, p. 194].

- *Hierarchy*: "Hierarchies can accomplish complex organizational tasks, but they are often associated with organizational properties inimical to innovation, such as slow (bureaucratic) decision making and weak incentives." A strong hierarchy is expected to hamper innovation because heavy organizational structures become barriers to the circulation of new ideas, communication flows, and decision-making within the firm [Teece, 1996, p. 200].
- *Bureaucratic decision-making*: It seems logical to think that agile decision processes favor innovation, while the decision processes linked to very rigid and slow structures clearly jeopardize the innovation process. The author argues that "the sharpening of global competition, and diversification (organizationally and geographically) in the sources of new knowledge economy compels firms to make decisions faster, and to reduce time to market in order to capture value from technological innovation. It seems clear that to accomplish such responsiveness, organizations need new structures and different decision-making protocols to facilitate entrepreneurial and innovative behaviour." However, this is not

always the case, and there are organizations where the processes and decisions are clearly bureaucratic and slow, acting as a barrier to innovation. If this is the case, strong leadership can overcome these organizational barriers, but this kind of leadership is not always present in the firm. Thus, in order to improve responsiveness of the firm and meet market requirements, it is necessary to achieve better and faster decision-making processes [Teece, 1996, p. 201]. To do so, as cited earlier, Burgelman [1984] and Burgelman and Sayles [1988] proposed a series of actions and measures such as special business units, new ventures department, new venture divisions, and independent business units. Therefore, it is necessary to create cross-functional structures that are supported by the corporate culture for its development.

- *Low-powered incentives*: Williamson [1985, cited in Teece, 1996] refers to "low-powered incentives" as those where the covariance between employee compensation and business performance is low. This refers to the difficulty in establishing incentive systems that favor innovation, usually through a grant of stock option. The employee's behavior balances between the risk-taking attitude associated with the innovation process and the safety feeling that is pursued by individuals. Thus, companies with high-risk environments, such as those related to innovation, have to compensate risk assumption with incentives to employees in order to make innovation attractive to employees.
- *Principal-agent distortions*: Usually large companies are managed by professional managers, who are evaluated by short-term results rather than by the long-term vision required for enhancing innovation. In addition, managers are completely at odds with the requirements of innovation, which are more related to risk taking and seeking results in the long run. In the same vein, we can refer to the work of Dyer *et al.* [2011], who differentiated between discovering and execution skills in order to explain the role of leadership in innovation. Thus, discovering-oriented managers are more likely to take risks, encourage innovation, and embrace risk in the company. However, execution-oriented managers aim only to accomplish goals and objectives, and they expect to be evaluated for ensuring the efficacy and efficiency of the company.
- *Myopia*: Organizations can remain closed to changes in the market, fueled by the short-term results that can underpin their mentality.

Accordingly, we can add the reference of Christensen [1997], who in his book *The Innovator's Dilemma* explains why successful companies fail to embrace disrupting technologies to create new markets and services. The author argues that managers prefer to focus their effort and resources on present customers' needs that can be fulfilled with current and feasible technologies rather than finding new ways of serving customers' needs through new, disrupting technologies.

- *Scope*: Teece [1996] referred to multi-product companies as firms that could more easily develop and commercialize "fusion technologies." These fusion technologies are the result of combining technological capabilities that support different business lines. The organization and structure of this kind of company respond to the necessity of integrating dissimilar product–technology environments.
- *Vertical integration*: Vertical integration in the industry favors innovation, as it allows systemic innovation, beyond stand-alone innovation. Integrating the value chain implies a higher inter-relation among the different core and support activities of the company. Consequently, it can facilitate the appearance of new links in the value chain, which may be the origin of process or product innovations.
- *Organizational culture and values*: "With respect to development, these include: the autonomy to try and fail; the right of employees to challenge the status quo; open communication to customers, to external sources of technology, and within the firm itself" [Teece, 1996, p. 206]. We will go deeply into this topic later on in this chapter, studying the role of corporate culture in innovation.
- *External linkages*: Companies look for alliances for the development and commercialization of new technologies. These agreements contain more structure, continuous interaction between the parties, open channels of information, greater confidence, and more orientation toward negotiation than to hierarchy. Managers have to learn to handle situations that go beyond their boundaries and involve mutual dependence.
- *Assessment*: "Economic research needs to pay greater attention to organizational structure, both formal and informal, and organizational research needs to understand the importance of market structure, internal structure, and the business environment" [Teece, 1996, p. 207].

Table 5.3. Types or Organizational Structures for Innovation.

Types or organizational structures for innovation	Description
Multi-product, integrated, hierarchical firms.	Bureaucratic decisions, absence of change culture, and high-powered incentives. Internally focused with low orientation toward the market. Decision-making is slow and ponderous. Innovation is possible if these organizations achieve a "breakout" [Downs, 1967] where a new division or venture team is set up, establishing autonomous strategic behavior. Large firms can easily support innovation and face uncertainty, set standards, and appropriability benefits using the technology embodied in products. Such firms are able to set up strategic alliances in order to harness entrepreneurial structures to promote new products.
High-flex "Silicon Valley"-type firms	Strong change culture, shallow hierarchies and high autonomy, simple and informal decision-making processes, open and quick communication and coordination, and lack of intellectual protection. They overcome hierarchical barriers and functional specialization. The lack of cash flow could be a barrier that can be solved with links with the venture capitalist community.
Virtual corporations	Absence of in-house manufacturing capabilities and vertical integration. These organizations have shallow hierarchies, innovative cultures, creative environments; are the first movers to the market; and have external linkages with manufacturers. The main problem is structural instability, where the innovator has problems capturing value and the manufacturer is likely the first competitor of the firm. Another hazard is related to R&D, since this function is linked to the manufacturer, or marketing activities that need the link between R&D and user feedback. All this means that coordinating costs rise and that more integrated structures are required to respond to the market, substituting or displacing virtual structures.

(*Continued*)

Table 5.3. (*Continued*)

Types or organizational structures for innovation	Description
Conglomerates	This is not exactly a structural form. It is decentralized, uses internal capital to fund technologies, venture capital is absent, headquarters managers act as external capital agents, and different corporate cultures are appointed. Accordingly, there are difficulties in building a change culture, not being the instinctive structure for markets with rapid technological change.
Alliance enterprise	"We define an alliance enterprise as a virtual corporation that has developed strong commitments to other enterprises, usually through equity-based links to affiliated enterprises lying upstream, downstream, horizontal, and lateral from its core business."

Source: Teece [1996, p. 216].

Teece [1996] argues that the management of the firm requires it to be aware of at least the firm's boundaries, internal formal structure, internal informal structure, and external linkages. However, instead of analyzing all the possible combinations of these variables, the author depicts a characterization of different types of organizations, distinguishing between stolid, multi-product, integrated hierarchies; high-flex "Silicon Valley"-type firms; hollow corporations of various types; and conglomerates of various types. In Table 5.3, we provide a summary of the different types of organizational structures according to the author.

Teece [1996, p. 216] recognizes the difficulty of matching types of innovation and organizational structures. However, the author argues that it is possible to match organizational forms with the existing capabilities of the firm and the types of innovation. The author states that, "As the interdependence between technologies increases, pure market forms are less effective at achieving the requisite coordination. The more systemic the innovation, the greater the interdependence." Hence, the author depicts a categorization of structural forms depending on the nature of the innovation and capabilities, as shown in Table 5.4.

Table 5.4. A proposed Matrix of Innovation, Capabilities, and Preferred Organizational Forms.

	Types of innovation	
	Autonomous	**Systemic**
Capabilities existing in house	Silicon Valley-type	Multi-product integrated
Capabilities existing outside	Virtual (outsourcing everything and anything)	Alliances (virtual with equity)
Capabilities must be created	Alliances (virtual with equity) Silicon Valley-type	Silicon Valley-type

Source: Teece [1996].

5.2 LEADING INNOVATION AND CHANGE

5.2.1 *The Relationship Between Leadership and Innovation*

Leadership is a process conceived as a set of actions conducive to achieve the purposes agreed to by the members of the organization. Leadership can also be defined as influence, that is, the art or process of influencing people to make them voluntarily meet group goals [Koont and Weihrich, 1998], where the source of the influence can be formal or informal [Robbins and Coulter, 2013]. Therefore, following this definition, the hypothesis that leadership can influence innovation implies that there is a series of actions undertaken by the leader that can trigger innovation or create the conditions within the organization to ease the innovation stream.

The relationship between leadership and innovation has been widely studied [Manz *et al.*, 1989; Mumford *et al.*, 2002; Nemanich and Vera, 2009; Jansen *et al.*, 2009] finding that leadership is one of the most clear determinants of innovation. Horth and Vehar [2015, p. 2] posited that "leadership is the most important factor needed to foster creativity and fuel innovation at the individual, team, and organizational levels. Leaders must act in ways that promote and support innovation in their culture." A corporate culture that enhances innovation is one of the main drivers of innovation in the firm, by creating the conditions to ease the team and individuals to promote innovation. Innovation culture is addressed in a later section of this chapter.

There is strong evidence in the literature that leadership is important for innovation [Nadler and Tushman, 1990; Denti and Hemlin, 2012], reporting also a decisive role of leadership in enhancing creativity [Mumford *et al.*, 2002; Tierney *et al.*, 1999; Amabile *et al.*, 2004]. Somech [2006] argues that corporate leaders are the key drivers who trigger or inhibit innovation in their organizations, and the lack of leadership can lead to deficient innovation processes [West *et al.*, 2003]. Leadership behaviors and activities such as practicing charismatic leadership, openness to new ideas and intellectual stimulation, recognizing individual contributions, and giving feedback as a way of contingent reward are all practices that have been reported to stimulate creativity and innovativeness [Amabile, 1997].

Leadership for innovation is not a homogeneous approach. Leaders can adopt different roles in their relationships with innovation. Thus, the leader may take the role of inventor (promoting the use of technological know-how to be used in the development of new products and services), entrepreneur (implementing innovation strategies in the organization), gatekeeper (gathering and processing information in the organization and its environment), champion (promoting the organizational adoption of innovations), and sponsor (promoting innovation through their hierarchical position) [Schumpeter, 1939; Nam and Tatum, 1997; Shane *et al.*, 1994; Hauschildt and Kirchmann, 2001; Roberts and Fusfeld, 1981].

Barsh *et al.* [2008] maintain that while senior executives cite innovation as an important driver of growth, few of them explicitly report managing it. According to the authors' research, about one-third of the interviewed leaders declare that they manage innovation on an *ad hoc* basis only when necessary, and another third declare that they manage it as part of the leadership agenda. Moreover, the authors found that most senior leaders do not actively encourage and model innovation behavior.

Barsh *et al.* [2008] suggest a number of practical steps to advance innovation. First, leadership must define the kind of innovation that drives growth and helps to meet strategic objectives, since the absence of that direction pushes the business to promote only incremental innovation. Second, it is necessary to add innovation to the formal agenda and regular leadership meetings and other forms of communication. Finally, the authors recognize the importance of setting performance metrics and targets for innovation (i.e., percentage of sales that come from products launched in the last 3 years).

5.2.2 *Leadership Styles and Innovation*

Even though leadership seems to be directly related to innovation, the relationship between them is not always positive. For explaining the latter, it is relevant to refer to the different kinds of leadership, so that we can explain the effect of each style on innovation.

First, we start by making reference to the concept of "participative leadership." This kind of leadership aims to increase the participation of individuals in the organization, share decision-making power and problem solving and provide consultation within the organization [Bass and Stogdill, 1990]. The key point in this kind of leadership is the existence of the joint decision-making processes or at least shared influence by a leader and followers [Koopman and Wierdsma, 1998]. This kind of leadership is seen to be positively correlated with innovation, since high levels of participation, consultation, and joint decision-making will increase the engagement and active participation of individuals in the organization. Accordingly, Burpitt and Bigoness [1997] claim that participative leaders achieve high rates of innovation in teams by giving them the freedom to develop new solutions and promoting their participation in projects. Participative leadership triggers creativity and the development of new ideas [Nijstad *et al.*, 2002], but may raise the level of conflict during the innovation period [Yan, 2011].

Second, we present the concept of "directive leadership," which aims at guiding individuals in the solution of the problems of the organization by obedience to the orders and direction facilitated by the leader [Bass *et al.*, 1975]. This kind of leadership provides the team members with a clear framework for decision-making visibly in line with the leader's vision [Fiedler, 1989; Sagie, 1997]. According to Kanter [1981], directive leaders drive innovation through control, hierarchy, training, and monitoring. For this kind of leadership, it is particularly beneficial to set up clear rules [Somech, 2006]. Both directive and participative leadership can be associated with high levels of team outcomes and high performance [Sagie *et al.*, 2002; Katzenbach and Smith, 1993].

In line with what we referred to earlier as participative leadership and going one step further, Rosener [1990] introduced the concept of "interactive leadership." Interactive leadership is characterized by the encouragement of participation by individuals, high levels of information and power sharing, the empowerment of employees for different tasks, and efforts to enhance employee feelings of self-worth. According to Bossink [2004], interactive leadership facilitates innovation because the leader empowers innovators within the organization to innovate, stimulating and allowing employees to innovate, cooperating with innovative employees, and teaching others how to be innovation leaders in the organization. At the same time, interactive leadership encourages employees to participate and contribute, having a positive impact on the innovation climate of the organization. Thus, interactive leadership involves guidance and support to employees on how to innovate [Bossink, 2007].

Another leadership style is the so-called "charismatic leadership." The first scholar who referred to this term was Weber [1947], followed by the contributions of House [1977] and Conger and Kanungo [1987], who explained that charismatic leadership is characterized by four key features: formulating and owning a vision, making risky assumptions to achieve the vision, being aware of the followers' needs, and demonstrating original behavior. According to Shamir *et al.* [1993, p. 577], charismatic leaders are those who "transform the needs, values, preferences, and aspirations of followers from self-interests to collective interests. Further, they cause followers to become highly committed to the leader's mission, to make significant personal sacrifices in the interest of the mission, and to perform above and beyond the call of duty." Charismatic leadership entices followers because they are able to envisage a promising future rather than causing discontent with the status quo [Nadler and Tushman, 1990; Pawar and Eastman, 1997]. There is evidence that charismatic leadership increases commitment, generates energy, and drives individuals toward new objectives and values [Nadler and Tushman, 1990; James and Lahti, 2011]. In relation to the promotion of innovation, some authors have reported that charismatic leaders create respect and loyalty, and a collective sense of mission leadership and team innovativeness [Eisenbach *et al.*, 1999]. Paulsen *et al.* [2009] posit that "leaders who are more transformational in style influence followers by affecting their sense of identity. At the same time, this sense of identity influences how well teams adopt and follow a cooperative strategy to resolve issues and make decisions" [Paulsen *et al.*, 2009, p. 2]. However, the existence of charismatic leadership alone is not a sufficient condition to guarantee innovation, since charismatic leadership has to be complemented by other leadership qualities in order to ensure organizational transformation [Bass, 1985; Nadler and Tushman, 1990].

We want to further make a distinction between transformational and transactional leadership. Both transformational leadership and transactional leadership have emerged together [Burns, 1978]. First, we refer to the concept of "transformational leadership," which was primarily introduced by Burns [1978] and further developed by Bass [1999], who defined it as a kind of leadership that moves followers beyond self-interests through a set of actions such as charisma, inspiration, intellectual stimulation, or individualized consideration. Transformational leadership is also the most actively researched leadership style with regard to innovation and change [Kesting *et al.*, 2016]. There has been some debate about the similarities and differences between charismatic and transformational leadership styles. House, who is considered the father of charismatic leadership, assumes very little differences between both the concepts. However, other authors such as Bass and Avolio [1994] affirm that transformational leadership has a broader view than charismatic leadership.

Transformational leadership consists of the following three factors: a charismatic leadership that inspires subordinates, provides individualized consideration to the employees thus giving support and paying attention individually, and provides intellectual stimulation to promote creativity [Howell and Avolio, 1993; Bycio *et al.*, 1995; Koh *et al.*, 1995]. According to Bass [2008], transformational leadership entails individualized consideration (addressing individual's needs, acting as a coach or mentor, and celebrating individual contributions), intellectual stimulation (encouraging new ideas, independent thinking, and challenging assumptions), inspirational motivation (articulating a vision and inspiring others to pursue it), and idealized influence (providing a model for high ethical behavior, respect, and trust). The authors state that "followers want to identify with such leadership. Intellectual stimulation is displayed when the leader helps followers become more innovative and creative. Individualized consideration is displayed when leaders pay attention to the developmental needs of followers and support and coach the development of their followers" [Bass, 1999, p. 11], making it clear that transformational leadership is related to innovation, since transformational leadership enhances motivation and pushes people to challenge the status quo [Keller, 2006]. In relation to innovation, Howell and Higgins [1990] determine a relationship between transformational leaders and champions, being able to take risks, with extraordinary personalities and the ability of envisioning and motivating others. Den Hartog *et al.* [1996] demonstrated that departments with a transformational leader primarily emphasize the values of support and innovation.

Rosing *et al.* [2011] and Kesting *et al.* [2016] reviewed the literature, analyzing the relationship between transformational leadership and innovation. Thus, in Table 5.5 you can see the most relevant findings.

Now we will address the concept of transactional leadership. This is focused on an exchange-based relationship that consists of two elements: a contingent reward and managing subordinates by correcting failures [Howell and Avolio, 1993; Bycio *et al.*, 1995; Daft, 1999], while the remaining focused on the follower's needs. Therefore, goal setting and a rewards system are important to achieve the proposed objectives [Bass, 1999]. Researchers have shown evidence of how transactional leadership has been applied to innovation projects. Howell and Avolio [1993, cited in Kesting *et al.*, 2016] argue that transactional leadership is more suitable for the implementation phase to keep things going and less appropriate for the stimulation of new ideas [Pieterse *et al.*, 2010, p. 611]. "Transactional leadership can be argued to be negatively related to innovative behavior because it is focused more on in-role performance and less on the stimulation of novel activities (which may be particularly detrimental for jobs where innovation is not an explicit part of the job description)." Also, Kesting *et al.* [2016] citing the work of Keller [1992] and Sillince [1994] affirmed that transactional leadership is suitable for incremental innovations and best suited to product innovations and R&D teams, while transformational leadership is better oriented to radical innovations.

Another interesting orientation in the study of the relationship between leadership and innovation is the view of strategic leadership. First, the concept of strategic leadership was initially studied by Hambrick and

Table 5.5. Relationship Between Transformational Leadership and Innovation.

Analyzed variables	Authors
High-diversity teams are correlated with innovation.	Kearney and Gebert [2009] and Shin and Zhou [2003]
Transformational leadership is effective for R&D projects.	Keller [1992] and [2006]
Positive relationship between transformational leadership and exploratory innovation.	Jansen *et al.* [2009]
Relationship between transformational leadership and organizational innovation is contingent with firm characteristics such as high climate of support for innovation, low centralization and formalization, as well as environmental features such as high uncertainty and competition.	Jung *et al.* [2008]
Raises followers' performance expectations, transforms their personal values and self-concepts, and moves them to a higher level of needs and aspirations	Jung *et al.* [2003] and Kahai *et al.* [2003]
Relationship between transformational leadership and innovation in firms with high organizational learning.	Garcia-Morales *et al.* [2008]
Transformational leadership is related to innovation when climate for excellence is also high.	Eisenbeiss *et al.* [2008]
Transformational leadership increases self-efficacy, raises intrinsic motivation, and contributes to employees' psychological empowerment.	Gumusluoğlu and Ilsev [2009] and Paulsen *et al.* [2009]
A positive relationship between transformational leadership and organizational innovation. Results of the analysis revealed that transformational leadership had a significant positive effect on organizational innovation.	Gumusluoğlu and Ilsev [2009]
Influences followers' attitudes optimistically and creates an overall positive culture.	McColl-Kennedy and Anderson [2002]
Transformational leadership could increase the level of trust.	Dirks and Ferrin [2002]
Transformational leadership behaviors have been found to have strong positive effects on the levels of innovation, risk taking, and creativity within business units.	Howell and Avolio [1993]

Source: Based on the work of Rosing *et al.* [2011] and Kesting *et al.* [2016].

Mason [1984], referring to it as the "upper echelons perspective" and studying organizational processes and outcomes. The analysis of strategic leadership focuses on executives who have overall responsibility for an organization [Finkelstein and Hambrick, 1996]. The relationship between strategic leadership and innovation has been studied by Cooper and Schendel [1976], who argued that executive decisions regarding innovation had important strategic implications. In the same vein, authors like Drucker [1985], Ireland and Hitt [1999], Bossink [2004], and Makri and Scandura [2010] have defended the importance of strategic decision-makers and their hierarchical position and the necessity of top management support and involvement to promote innovation. Strategic leadership uses the top position of the firm "to anticipate, envision, maintain flexibility, think strategically, and work with others to initiate changes that will create a viable future for the organization" [Ireland and Hitt, 1999, p. 63]. The literature has shown how members of the top management play a critical role in promoting innovation processes in organizations [Hansen and Kahnweiler, 1997; Papadakis and Bourantas, 1998].

Strategic leadership works in two areas. On the one hand, strategic leaders establish organizational structures, processes, and culture to promote innovation, which means creating the environmental conditions for that end [Michaelis *et al.*, 2009]. On the other hand, strategic leaders also play a role in the generation of new ideas, conceptualization, new product development [Kam Sing Wong, 2013], and discussing technical and design issues [Nam and Tatum, 1989].

Michaelis *et al.* [2009, p. 411] found that "trust in top management has a stronger indirect effect through affective commitment to change on innovation implementation behaviour than charismatic leadership. This result indicates that both sentiments regarding top management and immediate managers are important and complementary for successful innovation implementation."

Elenkov *et al.* [2005] posit a list of four activities carried out by strategic leaders that can be supportive for innovation. These activities are described in Table 5.6.

Finally, strategic leadership also studies the importance of personal traits required to become a successful leader. Thus, traits like extraversion, openness to experience, agreeableness, conscientiousness, and neuroticism

Table 5.6. Different Actions to Promote Innovation Form Strategic Leadership.

Actions to promote innovation	Authors
Due to their position in the firm, strategic leaders are able to find the main trends affecting the company and providing this information to the rest of the organization, which leads to higher levels of innovation.	Papadakis and Bourantas [1998]
Strategic leaders can present an exciting vision of the future to the organization, which can be achieved through innovation activity.	Hansen and Kahnweiler [1997]
Strategic leaders can also ease organizational innovation by supporting change champions within the firm.	Kanter [1985]
Creating an organizational culture that encourages productivity by means of rewards and enhancement of productive relationships.	Shamir *et al.* [1993], Podsakoff *et al.* [1996], and Avolio [1999]

Source: Elenkov *et al.* [2005, p. 669].

have shown a correlation to leadership behaviors [Judge and Bono, 2000; Judge *et al.*, 2002]. Others have identified anticipating, envisioning, flexibility, working with others, and strategic thinking as the main traits [Elenkov *et al.*, 2005].

Another interesting approach to the study of leadership and innovation stems from the perspective of the capacities required by the leader to promote innovation. It is interesting to cite the work of March [1991], who distinguished between exploration and exploitation abilities as two ways to promote organizational learning, which are crucial for organizational innovation [Rosing *et al.*, 2011]. This two-fold perspective is used in the work of Dyer *et al.* [2011], who, in their book *The Innovator's DNA*, depict a five-skill framework for innovation leaders. The five skills that compose the "innovator's DNA" are associating, questioning, observing, experimenting, and networking. Here is a short explanation of each concept:

- Association implies putting together diverse concepts. It is a way of spawning new ideas by generating links between unconnected ideas, places, concepts, and products.
- Questioning is the way in which a leader opens new possibilities for discovering new information and different points of view. The skill of

asking powerful questions enhances individuals' participation and promotes communication between leaders and followers.

- Observing is the ability to focus attention on people, situations, and places and getting relevant information from it. Observing may be referred to people in different work situations or witnessing processes, businesses, and technologies trying to find suitable new solutions.
- Experimenting is one of the key elements of the innovation process itself. The innovation stream usually starts with inspiration, followed by ideation and ending with experimentation. Experimentation implies putting into practice the ideation phase to check if new ideas are working or not.
- Networking is the ability to get connected with dissimilar people, environments, knowledge, places, products, or markets. The more diverse the networking, the more possibilities of generating new ways of thinking and new ideas.

In short, leadership is one of the key elements in the development of innovation within the organization. However, we cannot say that there is only one type of leadership that is more effective for innovation, though there are different types of leadership that influence the development of innovation in different ways. However, it seems that there is a consensus among the different authors when affirming that openness, communication, participation, shared decision-making processes, and individual recognition are elements that clearly encourage innovation within the company.

5.3 CREATING THE BUSINESS CULTURE TO BACK UP INNOVATION

The concept of corporate culture has been broadly studied by a variety of authors and their approaches [Trice and Beyer, 1993; Schultz, 1995; Deal and Kennedy, 1999; Ashkanasy *et al.*, 2000; Martin, 2004]. Organizational culture can be defined as "a pattern of shared basic assumptions that was learned by a group as it solved its problems of external adaptation and internal integration, that has worked well enough to be considered valid and, therefore, to be taught to new members as the correct way to perceive, think, and feel in relation to those problems" [Schein, 1982, p.17].

It seems clear that corporate culture is key in triggering innovation within the organization and it is considered one of the drivers that can fuel innovative behavior among members the most in the organization. On many occasions, corporate culture is identified as a critical success factor for the company. Successful organizations have the ability to embed innovation into the organizational culture and management procedures [Syrett and Lammiman, 1997]. As well, some approaches have considered the importance of a corporation's culture to support innovation and its relevance for being a source of competitive advantage within the company.

Traditionally, the literature has analyzed culture variables that positively influence innovation [Ahmed, 1998; Martins and Terblanche, 2003; McLean, 2005; Mumford, 2000]. For instance, general variables such as the existence of an innovation culture [Gumusluoğlu and Ilsev, 2009] or supportive culture [Abbey and Dickson, 1983; Berson *et al.*, 2008; Wei and Morgan, 2004] have been pointed out as being helpful for innovation within the organization. For example, supportive cultures and organizational encouragement increase creativity and the number of new ideas from employees, since they have a greater feeling of emotional security [Amabile *et al.*, 1996; Baer and Frese, 2003]. Literature contends that corporate cultures that are supportive of innovation stem from values, which are the beliefs supporting daily actions [Frohman, 1998], serving as a base to encourage or hinder process innovation and performance [Detert *et al.*, 2000].

Furthermore, there is a set of more specific variables that are also reported to influence positively on innovation. Among these variables we can cite tolerance for failure as crucial to motivating innovation [Detert *et al.*, 2000]; participation in decision-making and organizational flexibility (e.g., low use of formal rules, broad job definitions, and flexible authority structure) [Hurley and Hult, 1998]; training to improve expertise among the workforce [Boothby *et al.*, 2010; Shipton *et al.*, 2006]; external focus or outside orientation, which indicates a disposition to get new information and allows idea generation and opportunity recognition [Atuahene-Gima and Murray, 2004]; future orientation and risk tolerance [Tellis *et al.*, 2009]; values and beliefs of corporate managers (e.g., open communication methods, open questioning, change support, and diversity) [Amabile, 1988; King and Anderson, 1990; Woodman *et al.*, 1993]; shared vision in the company strategy that may pursue innovation as a strategic goal [Lock and Kirkpatrick, 1995]; shared values like flexibility, autonomy, and cooperative teamwork as influencers of innovation and creativity, while rigidity, control, and order as dampers on creativity and innovation [Arad *et al.*, 1997]. It is also worth citing the work of Martins and Terblanche [2003], who presented a framework for the analysis of the relationship of corporate culture in innovation. These authors propose a framework that is composed of two layers. The first one refers to the different dimensions that shape corporate culture, such as strategy, customer focus, means to achieve objectives, management processes, employee needs and objectives, interpersonal relationships, and leadership. The second layer contains the determinants of corporate culture that influence creativity and innovation. These factors are organized in five areas, which include strategy, structure, support mechanisms, behavior that encourages innovation, and communication.

- First, according to Martins and Terblanche [2003], strategy can influence innovation through the determination of vision and mission of the company and purposefulness. The vision can address organizational efforts through innovation, even including creativity and innovation as a corporate objective or value. Many companies have established innovation as a shared value, which leads its activity and behavior within the firm.
- Second, the structure of the organization can influence innovation. This factor has been profoundly analyzed in Section 5.1, determining the importance of structure to ignite or hinder innovation and concluding that organic structures are more suitable for innovation. Concerning the work of Martins and Terblanche [2003], the authors identify flexibility (to make use of a job rotation program or to do away with formal and rigid job descriptions as determinants of innovation and creativity), freedom (decision-making and speed of decision), and cooperation (diversity and talented individuals, cross-functional teams, trust, and respect).
- Third, the authors argue that there are some supporting mechanisms that encourage innovation. Accordingly, reward and recognition (for thinking creatively, experimentation, and risk taking) and availability of resources (time, information technology, and creative people) are the main supportive instruments for encouraging innovation.

- Fourth, the authors identified some behaviors that enhance innovation such as mistake handling (mistake as a learning opportunity and mistake tolerance), idea generation (encouraging idea generation and fair idea evaluation), continuous learning culture (being inquisitive and keeping knowledge skills up to date), risk taking and experimentation, competitiveness based on knowledge, support for change (new ways of working and positive attitude toward change), and conflict handling.
- Finally, the authors posed open communication as one of the main determinants of an innovation culture.

We have discussed the main determinants of corporate culture in organizational innovation. This approach has an orientation from the corporate point of view. Yet, we can also take an approach from people, since a great part of the literature refers to innovation as an individualized phenomena [Sternberg and Lubart, 1999] and has been approached through the study of the individual personality, traits, abilities and experiences, and through processes [Williams and Yang, 1999]. Likewise, it is necessary to be aware of the importance of the influence of social environment on creative behavior [Amabile *et al.*, 1996]. In this sense, we will study how the different elements of culture can influence the perception and attitude of individuals toward innovation. In order to ease the innovation process within the organization, individuals require a supportive organizational environment toward work [Frese *et al.*, 1996]. Employees also need to be aware of problems and have the ability to act proactively, pointing to the importance of initiative at work [Parker, 1998]. Likewise, the work environment has to be supportive to people to make them feel safe at the time of taking interpersonal risks and worthy of individual contributions [Edmondson, 1999]. In the same vein, successful cooperation entails a caring climate in which employees feel safe being proactive in a social work environment [Emery *et al.*, 1996]. Baer and Frese [2003] found that both personal initiative and safety are determinants of process innovation.

One of the hypotheses is that corporate culture supports creativity within the firm, affecting the extent to which creativity is encouraged, sustained, and developed. A culture that supports innovation and creativity boosts new ways of approaching and overcoming problems, regarding creativity as normal, and depicting innovators as models to be matched [Lock and Kirkpatrick, 1995]. Being aware that corporate culture directly influences employee behavior, this can lead employees to feel more involved in the business and assume innovation as a corporate value [Dulaimi and Hartmann, 2006].

With the aim of measuring the environmental climate that supports creativity within an organization, Amabile *et al.* [1996] designed and validated an instrument for assessing the climate for creativity, called KEYS. These authors identified six support scales that differentiated between high-creativity and low-creativity organizational climates. On the one hand, the authors posit the following variables for creativity enhancement: organizational encouragement, supervisory encouragement, working group supports, freedom, sufficient resources, and challenge. On the other hand, workload pressure and organizational impediments are identified as obstacles that hamper creativity. In Table 5.7, we summarize the contributions of Amabile *et al.* [1996].

Table 5.7. Variables Enhancing Creativity in the Work Environment.

Variables enhancing creativity	**Measures**
Organizational encouragement	Encouragement of risk taking and idea generation Fair and supportive evaluation of new ideas Reward and recognition of creativity Collaborative idea flow across an organization and participative management and decision-making.
Supervisory encouragement	Goal clarity Open interactions between supervisor and subordinates Supervisory support of a team's work and ideas
Work group encouragement	Diversity in team members' backgrounds, mutual openness to ideas, constructive challenging of ideas, and shared commitment to the project.
Freedom/Autonomy	Creativity is fostered when individuals and teams have relatively high autonomy in the day-to-day conduct of the work and a sense of ownership and control over their own work and their own idea.
Resources	Resource allocation to projects is directly related to the projects' creativity levels.
Pressures	Distinguishing between excessive workload pressure and challenge; the first should have a negative influence on creativity, while the second is positive.

Source: Amabile *et al.* [1996, p. 1161].

It is also worth citing the work of Kanter [1988], who argues that innovation and creativity are more likely to occur within an organization that has integrative structures, multiple structural linkages inside and outside the organization, intersecting territories, and collective pride and faith in people's talents and emphasizes diversity, collaboration, and teamwork.

Tesluk *et al.* [1997] worked on how the organizational culture impacts creativity at the individual level. The authors identified five elements of organizational climate that impact creativity, such as goal emphasis, means emphasis, reward orientation, task support, and socio-emotional support.

REFERENCES

Abbey, A., and Dickson, J. W. [1983]. R&D work climate and innovation in semi-conductors, *Academy of Management Journal*, 26(2), pp. 362–368.

Ahmed, P. K. [1998]. Culture and climate for innovation, *European Journal of Innovation Management*, 1(1), pp. 30–43.

Aiken, M., Bacharach, S. B., and French, J. L. [1980]. Organizational structure, work process, and proposal making in administrative bureaucracies, *Academy of Management Journal*, 23(4), pp. 631–652.

Aldrich, H. [1999] *Organizations Evolving* (Sage Publishing).

Amabile, T. M. [1988]. A model of creativity and innovation in organizations, *Research in Organizational Behavior*, 10(1), pp. 123–167.

Amabile, T. M. [1997]. Motivating creativity in organizations: On doing what you love and loving what you do, *California Management Review*, 40(1), pp. 39–58.

Amabile, T. M., Conti, R., Coon, H., Lazenby, J., and Herron, M. [1996]. Assessing the work environment for creativity, *Academy of Management Journal*, 39(5), pp. 1154–1184.

Amabile, T. M., Schatzel, E. A., Moneta, G. B., and Kramer, S. J. [2004]. Leader behaviors and the work environment for creativity: Perceived leader support, *The Leadership Quarterly*, 15(1), pp. 5–32.

Arad, S., Hanson, M. A., and Schneider, R. J. [1997]. A framework for the study of relationships between organizational characteristics and organizational innovation, *The Journal of Creative Behavior*, 31(1), pp. 42–58.

Ashkanasy, N. M., Wilderom, C. P., and Peterson, M. F. [2000]. *Handbook of Organizational Culture and Climate* (Sage Publishing, New Delhi, India).

Atuahene-Gima, K., and Murray, J. Y. [2004]. Antecedents and outcomes of marketing strategy comprehensiveness, *Journal of Marketing*, 68(4), pp. 33–46.

Avolio, B. J. [1999]. *Full Leadership Development: Building the Vital Forces in Organizations* (Sage Publishing).

Baer, M., and Frese, M. [2003]. Innovation is not enough: Climates for initiative and psychological safety, process innovations, and firm performance, *Journal of Organizational Behavior*, 24(1), pp. 45–68.

Barsh, J., Capozzi, M. M., and Davidson, J. [2008]. Leadership and innovation, *McKinsey Quarterly*, 1, p. 36.

Bass, B. M. [1985]. *Leadership and Performance Beyond Expectations* (Collier Macmillan).

Bass, B. M. [1999]. Two decades of research and development in transformational leadership, *European Journal of Work and Organizational Psychology*, 8(1), pp. 9–32.

Bass, B. M. [2008]. *The Bass Handbook of Leadership: Theory. Research, and Managerial Applications* (Simon and Shuster, New York).

Bass, B. M., and Avolio, B. J. [1994]. *Improving Organizational Effectiveness Through Transformational Leadership* (Sage Publishing, Thousand Oaks, CA).

Bass, B. M., and Stogdill, R. M. [1990]. *Bass & Stogdill's Handbook of Leadership: Theory, Research, And Managerial Applications* (Simon and Schuster, New York).

Bass, B. M., Valenzi, E. R., Farrow, D. L., and Solomon, R. J. [1975]. Management styles associated with organizational, task, personal, and interpersonal contingencies, *Journal of Applied Psychology*, 60(6), p. 720.

Berson, Y., Oreg, S., and Dvir, T. [2008]. CEO values, organizational culture and firm outcomes, *Journal of Organizational Behavior*, 29(5), pp. 615–633.

Boothby, D., Dufour, A., and Tang, J. [2010]. Technology adoption, training and productivity performance, *Research Policy*, 39(5), pp. 650–661.

Bossink, B. A. [2004]. Effectiveness of innovation leadership styles: A manager's influence on ecological innovation in construction projects, *Construction Innovation*, 4(4), pp. 211–228.

Bossink, B. A. [2007]. Leadership for sustainable innovation, *International Journal of Technology Management & Sustainable Development*, 6(2), pp. 135–149.

Burgelman, R. A. [1984]. Designs for corporate entrepreneurship in established firms, *California Management Review*, 26(3), pp. 154–166.

Burgelman, R. A., and Sayles, L. R. [1988]. *Inside Corporate Innovation* (Simon and Schuster, New York).

Burns, J. M. [1978]. *Leadership* (Harper & Row, New York).

Burns, T., and Stalker, G. M. [1961]. *The Management of Innovation* (Tavistock Publications, London).

Burpitt, W. J., and Bigoness, W. J. [1997]. Leadership and innovation among teams: The impact of empowerment, *Small Group Research*, 28(3), pp. 414–423.

Bycio, P., Hackett, R. D., and Allen, J. S. [1995]. Further assessments of Bass's [1985] conceptualization of transactional and transformational leadership, *Journal of Applied Psychology*, 80(4), pp. 468–478.

Christensen, C. M. [1997]. *The Innovator's Dilemma: When New Technologies Cause Great Firms to Fail* (Harvard Business School Press, Boston, MA).

Conger, J. A., and Kanungo, R. N. [1987]. Toward a behavioral theory of charismatic leadership in organizational settings, *Academy of Management Review*, 12(4), pp. 637–647.

Cooper, A. C., and Schendel, D. [1976]. Strategic responses to technological threats, *Business Horizons*, 19(1), pp. 61–69.

Covin, J. G., and Slevin, D. P. [1989]. Strategic management of small firms in hostile and benign environments, *Strategic Management Journal*, 10(1), pp. 75–87.

Daft, R. L. [1999]. *Leadership: Theory and Practice* (Dryden Press, Fort Worth, TX).

Deal, T. E., & Kennedy, A. A. [1999]. *The New Corporate Cultures* (Perseus, New York).

Den Hartog, D. N., Van Muijen, J. J., and Koopman, P. L. [1996]. Linking transformational leadership and organizational culture, *Journal of Leadership Studies*, 3(4), pp. 68–83.

Denti, L., and Hemlin, S. [2012]. Leadership and innovation in organizations: A systematic review of factors that mediate or moderate the relationship, *International Journal of Innovation Management*, 16(03), p. 1240007.

Detert, J. R., Schroeder, R. G., and Mauriel, J. J. [2000]. A framework for linking culture and improvement initiatives in organizations, *Academy of management Review*, 25(4), pp. 850–863.

Dirks, K. T., and Ferrin, D. L. [2002]. *Trust in Leadership: Meta-Analytic Findings and Implications for Research and Practice*, American Psychological Association.

Downs, A. [1967]. *Inside Bureaucracy* (Little Brown, Boston).

Drucker, P. F. [1985]. The discipline of innovation, *Harvard Business Review*, 63(3), pp. 67–72.

Dulaimi, M., and Hartmann, A. [2006]. The role of organizational culture in motivating innovative behaviour in construction firms, *Construction Innovation*, 6(3), pp. 159–172.

Dyer, J., Gregersen, H., and Christensen, C. M. [2011]. *The Innovator's DNA: Mastering the Five Skills of Disruptive Innovators* (Harvard Business Press, Harvard).

Edmondson, A. [1999]. Psychological safety and learning behavior in work teams, *Administrative Science Quarterly*, 44(2), pp. 350–383.

Eisenbach, R., Watson, K., and Pillai, R. [1999]. Transformational leadership in the context of organizational change, *Journal of Organizational Change Management*, 12(2), pp. 80–89.

Eisenbeiss, S. A., van Knippenberg, D., and Boerner, S. [2008]. Transformational leadership and team innovation: Integrating team climate principles, *Journal of Applied Psychology*, 93(6), p. 1438.

Elenkov, D. S., Judge, W., and Wright, P. [2005]. Strategic leadership and executive innovation influence: An international multi-cluster comparative study, *Strategic Management Journal,* 26(7), pp. 665–682.

Emery, C. R., Summers, T. P., and Surak, J. G. [1996]. The role of organizational climate in the implementation of total quality management, *Journal of Managerial Issues*, pp. 484–496.

Fiedler, F. E. [1989]. The effective utilization of intellectual abilities and job-relevant knowledge in group performance: Cognitive resource theory and an agenda for the future, *Applied Psychology*, 38(3), pp. 289–304.

Finkelstein, S., and Hambrick, D. C. [1996]. *Strategic Leadership: Top Executives and Their Effects on Organizations* (South-Western Pub).

Frese, M., Kring, W., Soose, A., and Zempel, J. [1996]. Personal initiative at work: Differences between East and West Germany, *Academy of Management Journal,* 39(1), pp. 37–63.

Frohman, A. L. [1998]. Managers at work: Building a culture for innovation, *Research-Technology Management*, 41(2), pp. 9–12.

Garcia-Morales, V. J., Matias-Reche, F., and Hurtado-Torres, N. [2008]. Influence of transformational leadership on organizational innovation and performance depending on the level of organizational learning in the pharmaceutical sector, *Journal of Organizational Change Management*, 21(2), pp. 188–212.

Gumusluoğlu, L., and Ilsev, A. [2009]. Transformational leadership and organizational innovation: The roles of internal and external support for innovation, *Journal of Product Innovation Management*, 26(3), pp. 264–277.

Hambrick, D. C., and Mason, P. A. [1984]. Upper echelons: The organization as a reflection of its top managers, *Academy of Management Review*, 9(2), pp. 193–206.

Hansen, C. D., and Kahnweiler, W. M. [1997]. Executive managers: Cultural expectations through stories about work, *Journal of Applied Management Studies*, 6(2), p. 117.

Hauschildt, J., and Kirchmann, E. [2001]. Teamwork for innovation — The "troika" of promotors, R&D *Management*, 31(1), pp. 41–49.

Horth, D. M., and Vehar, J. [2015]. *Innovation: How Leadership Makes the Difference*, Center for Creative Leadership.

House, R. J. [1977] *A 1976 Theory of Charismatic Leadership*, Hunt, J., and Larson, L. L. (eds.), *Leadership: The Cutting Edge* (University Press, Carbondale, Southern Illinois) pp. 189–207.

Howell, J. M., and Higgins, C. A. [1990]. Champions of technological innovation, *Administrative Science Quarterly*, pp. 317–34

Howell, J. M., and Avolio, B. J. [1993]. Transformational leadership, transactional leadership, locus of control, and support for innovation: Key predictors of consolidated-business-unit performance, *Journal of Applied Psychology*, 78(6), pp. 891–902.

Hurley, R. F., and Hult, G. T. M. [1998]. Innovation, market orientation, and organizational learning: An integration and empirical examination, *The Journal of Marketing*, pp. 42–54.

Ireland, R. D., and Hitt, M. A. [1999]. Achieving and maintaining strategic competitiveness in the 21st century: The role of strategic leadership, *The Academy of Management Executive*, 13(1), pp. 43–57.

James, K., and Lahti, K. [2011]. Organizational vision and system influences on employee inspiration and organizational performance, *Creativity and Innovation Management*, 20(2), pp. 108–120.

Jansen, J. J. P., Vera, D., and Crossan, M. [2009]. Strategic leadership for exploration and exploitation: The moderating role of environmental dynamism, *The Leadership Quarterly*, 20(1), pp. 5–18.

Judge, T. A., and Bono, J. E. [2000]. Five-factor model of personality and transformational leadership, *Journal of Applied Psychology*, 85(5), pp. 751–765.

Judge, T. A., Bono, J. E., Ilies, R., and Gerhardt, M. W. [2002]. Personality and leadership: A qualitative and quantitative review, *Journal of Applied Psychology*, 87(4), pp. 765–780.

Jung, D. I., Chow, C., and Wu, A. [2003]. The role of transformational leadership in enhancing organizational innovation: Hypotheses and some preliminary findings, *The Leadership Quarterly*, 14(4), pp. 525–544.

Jung, D. D., Wu, A., and Chow, C. W. [2008]. Towards understanding the direct and indirect effects of CEOs' transformational leadership on firm innovation, *The Leadership Quarterly*, 19(5), pp. 582–594.

Kahai, S. S., Sosik, J. J., and Avolio, B. J. [2003]. Effects of leadership style, anonymity, and rewards on creativity-relevant processes and outcomes in an electronic meeting system context, *The Leadership Quarterly*, 14(4), pp. 499–524.

Kam Sing Wong, S. [2013]. The role of management involvement in innovation, *Management Decision*, 51(4), pp. 709–729.

Kanter, R. M. [1981]. The middle manager as innovator, *Harvard Business Review*, 60(4), pp. 95–105.

Kanter, R. [1985]. Supporting innovation and venture development in established companies, *Journal of Business Venturing*, 1(1), pp. 47–60.

Kanter, R. M. [1988]. *When a Thousand Flowers Bloom: Structural, Collective, and Social Conditions for Innovation in Organization* (Harvard University, Boston).

Katzenbach, J. R., and Smith, D. K. [1993]. *The Wisdom of Teams: Creating the High-Performance Organization* (Harvard Business Press, Boston).

Kearney, E., and Gebert, D. [2009]. Managing diversity and enhancing team outcomes: The promise of transformational leadership, *Journal of Applied Psychology*, 94(1), pp. 77–89.

Keller, R. T. [1992]. Transformational leadership and the performance of research and development project groups, *Journal of Management*, 18(3), pp. 489–501.

Keller, R. T. [2006]. Transformational leadership, initiating structure, and substitutes for leadership: A longitudinal study of research and development project team performance, *Journal of Applied Psychology*, 91(1), pp. 202–2010.

Kesting, P., Ulhøi, J. P., Song, L. J., and Niu, H. [2016]. The impact of leadership styles on innovation-a review, *Journal of Innovation Management*, 3(4), pp. 22–41.

King, N. and Anderson, N. [1990]. *Innovation in working groups*, West, M. A., and Farr, J. L. (eds.), *Innovation and Creativity at Work*, pp. 81–100.

Koh, W. L., Steers, R. M., and Terborg, J. R. [1995]. The effects of transformational leadership on teacher attitudes and student performance in Singapore, *Journal of Organizational Behavior*, 16(4), (Wiley, USA) pp. 319–333.

Koont, H., and Weihrich, H. [1998]. *Administración. Una Perspectiva Global* (McGraw-Hill, México).

Koopman, P. L., and Wierdsma, A. F. M. [1998]. Participative management. Personnel psychology, Drenth, P., Thierry, H., Willems, P. and Wolff (eds.), Chapter 3, *Handbook of Work and Organizational Psychology* (Wiley, New York) pp. 297–324.

Lawrence, P. R., and Lorsch, J. W. [1967]. Differentiation and integration in complex organizations, *Administrative Science Quarterly*, pp. 1–47.

Lock, E. A., and Kirkpatrick, S. A. [1995]. Promoting creativity in organizations, Ford, and Gioia, D. (eds.), *Creative Action in Organizations: Ivory Tower Visions and Real-world Voices* (Sage, London) pp. 115–120.

Makri, M., and Scandura, T. A. [2010]. Exploring the effects of creative CEO leadership on innovation in high-technology firms, *The Leadership Quarterly*, 21(1), pp. 75–88.

Manso, G. [2011]. Motivating innovation, *The Journal of Finance*, 66(5), pp. 1823–1860.

Manz, C. C., Bastien, D. T., Hostager, T. J., and Shapiro, G. L. [1989]. Leadership and innovation: A longitudinal process view, *Research on the Management of Innovation: The Minnesota Studies*, pp. 613–636.

March, J. G. [1991]. Exploration and exploitation in organizational learning, *Organization Science*, 2(1), pp. 71–87.

Martin, J. [2004]. *Organizational Culture* (Wiley Encyclopedia of Management).

Martins, E. C., and Terblanche, F. [2003]. Building organisational culture that stimulates creativity and innovation, *European Journal of Innovation Management*, 6(1), pp. 64–74.

McColl-Kennedy, J. R., and Anderson, R. D. [2002]. Impact of leadership style and emotions on subordinate performance, *The Leadership Quarterly*, 13(5), pp. 545–559.

McLean, L. D. [2005]. Organizational culture's influence on creativity and innovation: A review of the literature and implications for human resource development, *Advances in Developing Human Resources*, 7(2), pp. 226–246.

Michaelis, B., Stegmaier, R., and Sonntag, K. [2009]. Affective commitment to change and innovation implementation behavior: The role of charismatic leadership and employees' trust in top management, *Journal of Change Management*, 9(4), pp. 399–417.

Mintzberg, H. [1979]. *The Structuring of Organizations: A Synthesis of the Research* (Prentice-Hall).

Mintzberg, H. [1980]. Structure in 5's: A synthesis of the research on organization Design, *Management Science*, 26(3), pp. 322–341.

Mumford, M. D. [2000]. Managing creative people: Strategies and tactics for innovation, *Human Resource Management Review*, 10(3), pp. 313–351.

Mumford, M. D., Scott, G. M., Gaddis, B., and Strange, J. M. [2002]. Leading creative people: Orchestrating expertise and relationships, *The Leadership Quarterly*, 13(6), pp. 705–750.

Nadler, D. A., and Tushman, M. L. [1990]. Beyond the charismatic leader: Leadership and organizational change, *California Management Review*, 32(2), pp. 77–97.

Nam, C. H., and Tatum, C. B. [1997]. Leaders and champions for construction innovation, *Construction Management & Economics*, 15(3), pp. 259–270.

Nam, C. H., and Tatum, C. B. [1989]. Toward understanding of product innovation process in construction, *Journal of Construction Engineering and Management*, 115(4), pp. 517–534.

Nemanich, L. A., and Vera, D. [2009]. Transformational leadership and ambidexterity in the context of an acquisition, *The Leadership Quarterly*, 20(1), pp. 19–33.

Nijstad, B. A., Stroebe, W., and Lodewijkx, H. F. [2002]. Cognitive stimulation and interference in groups: Exposure effects in an idea generation task, *Journal of Experimental Social Psychology*, 38(6), pp. 535–544.

O'Reilly, C. A., and Tushman, M. L. [2004]. The ambidextrous organization, *Harvard Business Review*, 82(4), pp. 74–81.

Papadakis, V., and Bourantas, D. [1998]. The chief executive officer as corporate champion of technological innovation: An empirical investigation, *Technology Analysis & Strategic Management*, 10(1), pp. 89–110.

Parker, S. K. [1998]. Enhancing role breadth self-efficacy: The roles of job enrichment and other organizational interventions, *Journal of Applied Psychology*, 83(6), pp. 835–852.

Paulsen, N., Maldonado, D., Callan, V. J., and Ayoko, O. [2009]. Charismatic leadership, change and innovation in an R&D organization, *Journal of Organizational Change Management*, 22(5), pp. 511–523.

Pawar, B. S., and Eastman, K. K. [1997]. The nature and implications of contextual influences on transformational leadership: A conceptual examination, *Academy of Management Review*, 22(1), pp. 80–109.

Phene, A., Tallman, S., and Almeida, P. [2012]. When do acquisitions facilitate technological exploration and exploitation? *Journal of Management*, 38(3), pp. 753–783.

Pieterse, A. N., Van Knippenberg, D., Schippers, M., and Stam, D. [2010]. Transformational and transactional leadership and innovative behavior: The moderating role of psychological empowerment, *Journal of Organizational Behavior*, 31(4), pp. 609–623.

Podsakoff, P. M., MacKenzie, S. B., and Bommer, W. H. [1996]. Transformational leader behaviors and substitutes for leadership as determinants of employee satisfaction, commitment, trust, and organizational citizen, *Journal of Management*, 22(2), pp. 259–298.

Robbins, S. P., and Coulter, M. A. [2013]. *Management*, 12th edn. (Pearson, Boston).

Roberts, E. B., and Fusfeld, A. R. [1981]. Staffing the innovative technology-based organization, *Sloan Management Review*, 22(3), pp. 19–34.

Rosener, J. B. [1990]. Ways women lead, *Harvard Business Review*, 68(6), pp. 119–125.

Rosing, K., Frese, M., and Bausch, A. [2011]. Explaining the heterogeneity of the leadership-innovation relationship: Ambidextrous leadership, *The Leadership Quarterly*, 22(5), pp. 956–974.

Sagie, A. [1997]. Leader direction and employee participation in decision making: Contradictory or compatible practices? *Applied Psychology*, 46(4), pp. 387–415.

Sagie, A., Zaidman, N., Amichai-Hamburger, Y., Te'eni, D., and Schwartz, D. G. [2002]. An empirical assessment of the loose–tight leadership model: Quantitative and qualitative analyses, *Journal of Organizational Behavior*, 23(3), pp. 303–320.

Sarkees, M., and Hulland, J. [2009]. Innovation and efficiency: It is possible to have it all, *Business Horizons*, 52(1), pp. 45–55.

Schein, E. H. [1982]. *Organizational Culture and Leadership* (John Wiley & Sons).

Schultz, M. [1995]. *On Studying Organizational Cultures: Diagnosis and Understanding* (Walter de Gruyter).

Schumpeter, J. A. [1939]. *Business Cycles: A Theoretical, Historical, and Statistical Analysis of the Capitalist Process* (McGraw-Hill, New York).

Shamir, B., House, R. J., and Arthur, M. B. [1993]. The motivational effects of charismatic leadership: A self-concept based theory, *Organization Science*, 4(4), pp. 577–594.

Shane, S. A., Venkataraman, S., and Macmillan, I. C. [1994]. The effects of cultural differences on new technology championing behavior within firms, *The Journal of High Technology Management Research*, 5(2), pp. 163–181.

Shin, S. J., and Zhou, J. [2003]. Transformational leadership, conservation, and creativity: Evidence from Korea, *Academy of Management Journal*, 46(6), pp. 703–714.

Shipton, H., West, M. A., Dawson, J., Birdi, K., and Patterson, M. [2006]. HRM as a predictor of innovation: HRM as a predictor of innovation, *Human Resource Management Journal*, 16(1), pp. 3–27.

Sillince, J. A. [1994]. A management strategy for innovation and organizational design: The case of MRP2/JIT production management systems, *Behaviour & Information Technology*, 13(3), pp. 216–227.

Sine, W. D., Mitsuhashi, H., and Kirsch, D. A. [2006]. Revisiting Burns and Stalker: Formal structure and new venture performance in emerging economic sectors, *Academy of Management Journal*, 49(1), pp. 121–132.

Somech, A. [2006]. The effects of leadership style and team process on performance and innovation in functionally heterogeneous teams, *Journal of Management*, 32(1), pp. 132–157.

Sternberg, R. J., and Lubart, T. I. [1999]. The concept of creativity: Prospects and paradigms, *Handbook of Creativity*, 1, pp. 3–15.

Syrett, M., and Lammiman, J. [1997]. *From Leanness to Fitness. Developing Corporate Muscle* (Institute of Personnel and Development, London).

Taylor, F. T. [1911]. *The Principles of Scientific Management* (Harper & Brothers, New York).

Teece, D. J. [1996]. Firm organization, industrial structure, and technological innovation, *Journal of Economic Behavior & Organization*, 31(2), pp. 193–224.

Tellis, G. J., Prabhu, J. C., and Chandy, R. K. [2009]. Radical innovation across nations: The preeminence of corporate culture, *Journal of Marketing*, 73(1), pp. 3–23.

Tesluk, P. E., Farr, J. L., and Klein, S. R. [1997]. Influences of organizational culture and climate on individual creativity, *The Journal of Creative Behavior*, 31(1), pp. 27–41.

Thompson, J. D. [1967]. *Organizations in Action: Social Science Bases of Administrative Theory* (Transaction publishers, New Jersey).

Tierney, P., Farmer, S. M., and Graen, G. B. [1999]. An examination of leadership and employee creativity: The relevance of traits and relationships, *Personnel Psychology*, 52(3), pp. 591–620.

Trice, H. M., and Beyer, J. M. [1993]. *The Cultures of Work Organizations* (Prentice-Hall).

Tushman, M. L., and O'Reilly, C. A. [2002]. *Winning Through Innovations* (Harvard Business School Press, Boston).

Tushman, M., Smith, W. K., Wood, R. C., Westerman, G., and O'Reilly, C. [2010]. Organizational designs and innovation streams, *Industrial and Corporate Change*, 19(5), pp. 1331–1366.

Weber, M. [1947]. *The Theory of Economic and Social Organization.* Trans. Henderson, A. M., and Parsons, T. (Oxford University Press, New York).

Wei, Y. S., and Morgan, N. A. [2004]. Supportiveness of organizational climate, market orientation, and new product performance in Chinese firms, *Journal of Product Innovation Management*, 21(6), pp. 375–388.

West, M. A., Borrill, C. S., Dawson, J. F., Brodbeck, F., Shapiro, D. A., and Haward, B. [2003]. Leadership clarity and team innovation in health care, *The Leadership Quarterly*, 14(4), pp. 393–410.

Williams, W. M., and Yang, L. T. [1999]. *Organizational Creativity*, Sternberg, R. (ed.), Chapter 19, *Handbook of Creativity* (Cambridge University Press, Cambridge) pp. 373–391.

Williamson, O. E. [1985]. *The Economic Institutions of Capitalism: Firms, Markets, Relational Contracting*, Vol. 866 (Free Press, New York).

Woodman, R. W., Sawyer, J. E., and Griffin, R. W. [1993]. To ward a theory of organizational creativity, *Academy of Management Review*, 18(2), pp. 293–321.

Yan, J. [2011]. An empirical examination of the interactive effects of goal orientation, participative leadership and task conflict on innovation in small business, *Journal of Developmental Entrepreneurship*, 16(03), pp. 393–408.

Yang, H., and Atuahene-Gima, K. [2007]. Ambidexterity in product innovation management: The direct and contingent effects on product development performance, *Annual Meeting of the Academy of Management*, RGC, p. 32.

CHAPTER 6

HUMAN CAPITAL, CREATIVITY, AND LEARNING

6.1 IMPORTANCE OF HUMAN CAPITAL FOR INNOVATION

There are different approaches to defining innovation. Kimberly and Evanisko [1981] depicted a three-fold perspective of innovation. The first one focused on the development of new products and services; the second one considers innovation as a process with different stages and phases that can be managed (see Chapter 2 for further information); and the third considers innovation as an organizational capability, referring to it as innovativeness or innovative capability. This third approach stems from the ability of the organization to put into practice a sustainable development strategy based on leadership and an appetite for change along with the acceptance of risk and experimentation.

Innovators are able to develop sustainable competitive advantages that are really difficult to imitate by competitors, since they have been able to build a set of organizational capabilities that are robust and unique and adapted to the market's needs [Ulrich and Lake, 1991]. Teece [2010, p. 190] describes this as "the sensing, seizing, and reconfiguring skills that the business enterprise needs if it is to stay in synch with changing markets, and which enable it not just to stay alive, but to adapt to and itself shape the (changing) business environment." Accordingly, the key skills that the business needs to fulfill the necessity for continuous adaptation to the market's needs necessarily lean on the individuals of the firm, who hold the competences, skills, and abilities to support organizational change. Therefore, human capital is the foundation for developing this set of abilities, since the skill set necessary for embedding the innovation mind-set into the organization necessarily derives from the staff and is reflected throughout the company's culture.

Schultz [1981] linked human capital with innate and acquired skills, which are important and should be developed through further investment. Consequently, Schultz emphasizes the importance of people in economic development and well-being, arguing that, "increases in the acquired abilities of people throughout the world and advances in useful knowledge hold the key to future economic productivity and to its contributions to human well-being ... Investment in population quality and in knowledge in large part determines the future prospects of mankind" [Schultz, 1981, p. xi].

At the organizational level, Bontis *et al.* [1999] defined human capital as the human factor in the organization, including the distinctive skills and expertise, the combined intelligence of the organization, and the human part of the organization qualified to learn, innovate, change, and deliver the creativity to ensure the long-term survival of the firm. Therefore, they make a distinction between the individual level, where the skills and expertise exist, and the organizational level, which combines the intelligence of the people and creates the basis for sustainable change. Human capital is also defined as a set of knowledge, capabilities, and skills held by employees, which, in the aggregate, constitute an important asset for the firm [Nahapiet and Ghoshal, 1998; Subramaniam and Youndt, 2005].

It is also necessary to cite the human capital theory advanced by Becker [1975], who approached individual decisions about investments in skills and knowledge with the aim of improving productivity, career choices, and other work characteristics, assuming that individuals choose their occupation or employment to maximize the present net economic and psychological value over their lifetimes. As Becker argues, "many workers increase their productivity by learning new skills and perfecting old ones while on the job. Presumably, future productivity can be improved only at a cost, for otherwise there would be an unlimited demand of training" [1975, p. 17]. Therefore, the author points out the importance of developing new skills and links it to an increase in productivity, but at the same time assumes that the impact on productivity comes at a cost.

Additionally, Grant [1995, p. 143] defines human capital as the "productive services that human beings offer to the firm in terms of their skills, knowledge, and reasoning and decision-making abilities"; Kamoche [1996, p. 216] refers to it as the "accumulated stock of knowledge, skills, and abilities that the individuals possess, which the firm has built up over time into an identifiable expertise"; and Barney [1995, p. 50] describes it as "all the experience, knowledge, judgment, risk-taking propensity, and wisdom of individuals associated with a firm."

Particularly, our approach to the study of human capital is made under the perspective of its link with innovation capacity of the firm. Consequently, we have studied and found substantial evidence in scientific literature correlating human capital and innovative capabilities in the firm. In this vein, Dakhli and De Clercq [2004] examined the effects of human and social capital on innovation at a geographical level. Consequently, using a country-level approach across 59 different nations, researchers found strong evidence of a positive relationship between human capital and innovation, while the relationship between social capital and innovation was only partially supported. In other relevant research, Subramaniam and Youndt [2005] examined the influence of intellectual capital (IC) on various innovative capabilities in organizations, and they found that human, organizational, and social capital and their interrelationships selectively influenced the incremental and radical innovative capabilities. More particularly, the researchers found that organizational capital positively influenced incremental innovative capability, while human capital interacted with social capital to positively influence radical innovative capability. Likewise, Marqués *et al.* [2006] conducted research on 222 Spanish firms in the biotech and telecommunication industries. They reached the conclusion that innovation competences positively affect the stock of intangibles and facilitate the development of relational capital, structural capital, and human capital, proving how innovation competences affect the stock of intangibles and facilitate the development of IC. Similarly, Thornhill [2006], who studied 845 Canadian manufacturing firms, concluded that a highly skilled workforce is most beneficial to firm performance in dynamic environments, while firms in stable manufacturing industries benefit more from investments in training. Moreover, the assumption that the knowledge within the firm is closely related to its products and services clarifies that the ability of the firm to produce new products and services depends on its organizational capabilities and human capital.

Although not all human capital resources are valuable for innovation, those who have differential, rare, or unique skills can constitute an added value to the firm. For instance, Lopez-Cabrales *et al.* [2006] investigated the role of core employees in firm competitiveness, testing the relationship between the value and uniqueness of core employees' knowledge, skills, and abilities (KSAs) and organizational capabilities that define the competitive advantage of the firm. The authors found a higher organizational capability for firms using the most valuable and unique core employees in comparison with those that do not. Also, it is worth citing the work of Laursen [2002], who studied a sample of 726 Danish firms with more than 50 employees in the manufacturing and private services. The results showed that human resource management (HRM) practices are more effective in influencing innovation performance when applied together, enhancing organizational complementarities. Additionally, the application of complementary HRM practices is more effective in knowledge-intensive industries.

It is also interesting to cite the work of Smith *et al.* [2005], who argued that development of human capital is critical for the development of new knowledge in firms, linking it with the new product development (NPD) processes. The authors conducted a field study of top management teams and knowledge workers from 72 technology firms, and they reached the conclusion that the organization's members' ability to combine and exchange new knowledge was determinant on the rate of introduction of new products and services. More particularly, the authors determined that the aggregate of existing knowledge of employees, the knowledge from member networks, and the organizational climate for risk-taking and teamwork is the basis of the firm's ability to innovate effectively.

6.2 CREATIVITY AND INNOVATION

Innovation is the way to materialize new knowledge and ideas in the market. Therefore, innovation necessarily leans on the development of new ideas and new and creative ways to solve problems or flaws. Therefore, innovation systematically stems from research and development aimed at finding new solutions or creating new products and services or new knowledge itself [Tidd *et al.*, 2009]. Likewise, innovation is often seen as the province of technical specialists in R&D, engineering, or design, where the underlying creative skills and problem-solving abilities are possessed by everyone [Tidd *et al.*, 2009]. Likewise, scholars have systematically studied the relationship between individual and team creativity with organizational innovation [Amabile, 1997].

Amabile [1997, p. 1] stated, "because creativity is the production of novel and appropriate (or useful) ideas, managers who wish to ensure the continued success of their businesses must find ways to encourage a continuing flow of creative ideas within their organizations. Those ideas will be needed not only for the development of new, valuable products and services, but also for the solution of ever-changing and ever more challenging business problems." Therefore, the authors not only recognize the importance of creativity for NPD but also for the ongoing evolution and survival of the firm.

Therefore, the first point is trying to differentiate between creativity and innovation, since creativity has usually been conceived as the first stage in the generation of novel ideas, whereas innovation is more focused on the implementation of these ideas at a later stage [Amabile, 1996]. However, many others point out that creativity does not occur in the early stages of the innovation process alone, but rather claim that innovation is a cyclical, iterative process of idea generation and implementation [Paulus *et al.*, 2012]. Some authors also relate creativity with a higher degree of novelty, whereas innovation may involve ideas that are relatively novel [Anderson *et al.*, 2004].

Following the literature, creativity has been defined under the output perspective approach, focusing on the production of ideas, products, processes, and procedures [Amabile, 1996; Oldham and Cummings, 1996]. The topic of creativity has been treated with the aim of explaining why some individuals, teams, or organizations are more likely than others to articulate novel ideas for processes, services, and products [Amabile, 1996].

Amabile [1998, p. 78] reasons that we usually identify creativity with the "arts and to think of it as the expression of highly original ideas."

However, the author defends creativity as a function of three components: "expertise (knowledge-technical, procedural, and intellectual), creative-thinking skills (determine how flexibly and imaginatively people approach problems), and motivation (an inner passion to solve the problem at hand leads to solutions far more creative than do external rewards, such as money)."

Farr and West [1990, p. 9] defined innovation as "the intentional introduction and application within a role, group, or organization of ideas, processes, products, or procedures, new to the relevant unit of adoption, designed to significantly benefit the individual, the group, the organization or wider society." This definition clearly makes a differentiation between creativity and innovation, since innovation implies the intentional introduction and application of new and improved ways of doing things; whereas creativity, on the other hand, refers to idea generation alone. Moreover, innovation could have impact at different levels such as individual, group, organization, or society, which is not necessarily the case for creativity.

As stated earlier, creativity is typically related to cultural products and services, such as visual arts, performances, and literature. Nevertheless, it can be noted that looking deeper into creativity during decision-making, it goes far beyond the production of cultural products since creativity can actually be expressed in many types of organizations and at many levels of hierarchy within them [Shalley and Gilson, 2004].

At the organizational level, creativity has become increasingly important in order to determine organizational performance, innovation, and long-term survival of the firm. Consequently, we could say that creativity is an important aspect of innovation, since innovation actually requires creativity [Shalley and Gilson, 2004]. Organizations try to exploit the ideas that emerge in the organizational context in an axiomatic process that lead the business to a distinctive source of competitive advantage [Anderson *et al.*, 2004]. Similarly, innovation is necessary for the overall competitiveness of the organization. The business needs to come up with new and better products, services, and processes to compete in the market arena. Businesses need to relentlessly provide products and services appropriate and needed at the corresponding time, and thus exercise creativity in their strategic and daily operations [Amabile, 1997].

Further, creativity is based on people within the organization who are the foundation for organizational creativity [Amabile, 1998], and this characteristic is linked to firm performance, organizational development, and survival of the firm [Nystrom, 1990].

As stated earlier, from the psychological perspective, creativity is referred to as a personality trait that is associated with individuals in the organization [Amabile, 1996; Barron, 1955]. Ford [1996] also attempted to explain under the psychological and sociological perspective how employees decide to act creatively. This individual perspective depicts a two-fold approach for employees, who have two options: being creative or undertake work routinely. Individuals in the organization will be influenced by their sense of making progress, their motivation, and the perception of their own knowledge and skills.

Focusing on the skills and abilities of individuals as a means to igniting creativity in the workplace, Woodman *et al.* [1993] proposed a model of individual creativity based on the interactionist model of creative behavior that is said to be a function of personality factors, cognitive style and ability, relevant task domain expertise, motivation, and social and contextual influences. Shalley and Gilson [2004] also refer to personality traits, creativity-relevant skills, and domain-specific knowledge as reflecting an individual's level of education, training, experience, and knowledge within a particular context. Likewise, the authors point out the importance of experience in a field as a necessary component for creative success because some level of expertise is needed to perform creative work. The authors add that creativity also needs relevant skills to be performed, defining them as "the ability to think creatively, generate alternatives, engage in divergent thinking, or suspend judgment. These skills are necessary because creativity requires a cognitive–perceptual style that involves the collection and application of diverse information, an accurate memory, use of effective heuristics, and the ability and inclination to engage in deep concentration for long periods of time" [Shalley and Gilson, 2004, p. 36].

However, there is an increasing need to understand how to nurture creativity in the organizational arena, finding the contextual factors that may boost or impede creativity within the organization. Shalley and Gilson [2004] referred to different elements that influence creativity in the workplace. They refer to leadership as one element that directly influences creativity; they state, "with regards to creative outcomes, managers play a key role in that they are often the individuals best suited to make the determination of whether an employee's outcome should be regarded as creative." They add, "in order for creativity to occur, leadership needs to play an active role in fostering, encouraging, and supporting creativity" [Shalley and Gilson, 2004, p. 35]. It is also worth citing the work of Mumford *et al.* [2002], who defend that leaders play an important role in igniting creativity, ensuring the right structure of the work environment, climate and culture, and human resource practices.

Tierney *et al.* [1999] studied 191 R&D employees from a large chemical company to test a multi-domain, interactionist creativity model of employee characteristics, leader characteristics, and leader–member exchange. The results showed that these factors and their interaction relate to employee creative performance as measured by supervisor ratings, invention disclosure forms, or research reports. Accordingly, Oldham and Cummings [1996] studied 171 employees from two manufacturing facilities, examining the independent and joint contributions of employees' creativity. They analyzed the relevant personal characteristics of employees and three characteristics of the organizational context, such as job complexity, supportive supervision, and controlling supervision. They measured creativity's performance using three indicators: patent disclosures written, contributions to an organization's suggestion program, and supervisory ratings of creativity. They found that employees produced the most creative work when they had appropriate creativity-relevant characteristics, worked on complex, challenging jobs, and were supervised in a supportive, non-controlling fashion. Therefore, the creative skills of individuals, meaningful work, and leadership become the most relevant determinants of creativity output.

Social and contextual factors are also relevant in explaining creativity in the workplace. Again, Shalley and Gilson [2004, p. 35] argue that, "if managers are aware of the important social and contextual factors at all levels, they should be better able to positively affect the occurrence of creativity." Hence, apart from individual characteristics, skills, or traits, Shalley and Gilson [2004] also mentioned job-level factors (job characteristics, role expectations and goals, sufficient resources, rewards, supervisory support, external evaluation of work), team or group factors (social context, group composition), and organizational-level factors (organizational climate, organizational-level human resource practices).

Creativity has special relevance from the team perspective. Gilson and Shalley [2004] developed a cluster analysis, finding that the more creative teams were those that perceived that their tasks required high levels of creativity, working on jobs with high task interdependence, were high on shared goals, valued participative problem solving, and had a climate supportive of creativity. In addition, the authors also claimed that members of the more creative teams spent more time socializing with each other and had moderate amounts of organizational tenure. More particularly, they argue, "it appears that if employees perceive that creativity is an important component of the job, whether they believed it was expected or whether it was specifically defined as part of the job, this may enhance an employee's motivation to try new things or link ideas from different areas. Thus, this perceived expectation for creativity seems to translate into more active engagement in creative processes" [Gilson and Shalley, 2004, p. 465]. Concerning task interdependence, the authors claim that motivation could explain this, since if others are dependent upon your work, being creative may be a way of helping each other.

Another approach to be considered is how employment relationships influence innovation. In a direct way, different authors have analyzed the impact that employment relationships have on innovation, be it all alone or through interaction. In this vein, Bornay-Barrachina *et al.* [2012] examined how employment relationships and human capital influence innovation. The authors used a sample of 150 innovative Spanish firms confirming that, while human capital favors innovation, employment relationships are not directly associated with innovation unless they take human capital into account. According to this analysis, human capital mediates the effects of some employment relationships (mutual investment) on innovation. Consequently, the study found that human capital affects a firm's innovation capability, demonstrating a positive and direct correlation between human capital and innovation. More deeply, the research showed the fundamental role played by human capital on innovation while considering both its direct and indirect effects. However, the authors focus on the fact that employment relationship models by themselves do not explain innovation; thus, it's necessary to consider other variables to better understand their relationship with innovation.

López-Cabrales *et al.* [2006] found that HR managers assessed their core employees as valuable human assets, finding an interesting conclusion that firms employing the most valuable and unique core employees attained the highest level of organizational capabilities in the sample, showing at the same time that this result was significant in the case of capabilities oriented to organizational culture, employee potential, strategic vision, and innovation.

On the other hand, employment relationships can also be directed to the creation of new KSAs that ultimately generate innovation. Thus, according to Lepak *et al.* [2003], who examined the relationships among four types of employment (knowledge-based, job-based, contract, and alliances) and firm performance, they found that a greater use of knowledge-based employment and contract work is positively associated with firm performance.

Following with the analysis of context or social factors that explain the development of creativity within the firm, we can cite the work of Martins and Terblanche [2003], who discussed the relationship between creativity, innovation, and culture. They found strategy, structure, support mechanisms, and behavior as determinants that encourage innovation and open communication. They also found that creativity and innovation were influenced by values, norms, and beliefs that can either support or inhibit creativity and innovation, depending on how they influence individual and group behavior.

6.3 IDEA MANAGEMENT: TOOLS TO ENHANCE CREATIVITY

If creativity is an important part of the innovation process, ideas are the materialization of the creative process itself and the fuel to spark innovation. Therefore, the process of idea generation is a cornerstone within the creativity practice. Consequently, idea management is important to promote innovation and firm development.

There are many authors who focus on the importance of ideas to ignite innovation at the organizational level. For instance, Van de Ven [1986, p. 1] defines innovation as "the development and implementation of new ideas by people who over time engage in transactions with others within an institutional order." Also, Boeddrich [2004, p. 274] argues that "[a]ll innovations originate from ideas, which are the results of the creative or rational thinking processes of employees, customers, suppliers, or universities, generated individually or in group sessions." Therefore, it is assumed, the greater the number of ideas a company generates, the richer the innovation results and firm performance [Boeddrich, 2004; Francis and Bessant, 2005].

It seems that innovation success depends on the number of ideas that the business is able to generate. It is further assumed that the pace of technological change, globalization, and business speed increase competition, and therefore, businesses need to constantly innovate to ensure long-term competitiveness. Consequently, organizations demand a continuous stream of ideas to ignite innovation [Björk and Magnusson, 2009].

For the purpose of this chapter, it is also worth defining what we mean by an idea. Koen *et al.* [2002, p. 7] define an idea as "the most embryonic form of a new product or service. It often consists of a high-level view of the solution envisioned for the problem identified by the opportunity," differentiating it from the term "concept" that is defined as "a well-defined form, including both a written and visual description, that includes its primary features and customer benefits combined with a broad understanding of the technology needed." Galbraith [1982, p. 7] also points out the unpredictable character of ideas and innovation in the organization. He argues that "the innovation in the story, like many others, is usually regarded as an accident. A combination of random events occurred to enable the engineer's ideas to be tested and developed."

Sometimes, there is a misunderstanding in the difference between the generation of new ideas and the process of idea management at the organizational level. To clarify this point, we cite the work of Mikelsone and Liela [2015] who differentiate between ideas and idea management. They argue that ideas represent the raw material for innovation, while idea management can be seen as the core of the innovation process. Schroeder *et al.* [2000, p. 108] also claim that "the process of innovation centers on the temporal sequence of activities that occur over time in developing and implementing new ideas from concept to concrete reality." This is perfectly logical in the sense that the more ideas generated, the higher the probability of selecting a very good one. Conversely, investigation has demonstrated that organizations do not operate idea selection efficiently [Mikelsone and Liela, 2015], the ideation process indeed being highly informal, knowledge-intensive, and erratic. These characteristics indicate that the outcome of the idea-generation process is highly uncertain and therefore, managing this phase of innovation is delicate and can turn out to be detrimental to a firm's innovation processes and ensuing performance unless it is strategically managed [van den Ende *et al.*, 2015].

For good reason, the early phase of innovation, in which ideas are born, is often considered or referred to as the "fuzzy front end" [van den Ende *et al.*, 2015]. There are some authors who point out the importance of shedding light on this fuzzy stage and do so by adopting management systems that can rationalize the ideation process within organizations. Thus, Boeddrich [2004] wrote, "according to decision-making theory, developing initial embryonic ideas into practicable project proposals reduces uncertainty. High quality, in terms of the initial decision for innovation projects has a positive influence on the quality of the innovation process in companies and increases the probability that successful products will result from this process" [Boeddrich, 2004, p. 274].

Accordingly, it is also worth citing the well-known study by Booz *et al.* [1982], which is based on knowledge accumulated from over 800 client projects and data acquired from just over 49 firms. It found that almost a third of all NPD project commercialization attempts were failures, independent of industry.

Coming back to the necessity of idea management, we refer to one of the seminal works conducted by Green *et al.* [1983], who studied the problems concomitant with the management of idea flows in the R&D lab. The authors proposed a basic four-stage process that goes from idea generation to idea capturing, idea retention, and idea retrieval. Also, Rowbotham *et al.* [1996] described a structured idea management approach, made up of seven stages: development of criteria, preparation for idea generation, idea selection, development of ideas, idea evaluation, ranking of ideas, and concept development. Likewise, El Bassiti and Ajhoun [2013] studied idea management as part of the whole innovation management process within the organization. The authors defined idea management as "the generation of new concepts, by combining organization's knowledge and collective intelligence, aligned by the organization's contextual factors [including strategy, goals, needs..." [El Bassiti and Ajhoun, 2013, p. 551]. The authors also depicted a model composed of four stages: generation, interlinking, improvement stage, and validation stage.

Björk and Magnusson [2009], in their investigation, "Where do good innovation ideas come from?" Stress the importance of the sources of innovation. They argue that ideas cannot depend only on a group of people within the firm or even a particular department, but, rather, it is a function that requires the participation of everyone either within the business or affiliated with it. The authors also declare that overall, the literature related to innovation idea generation and identification is wide, drawing on theories about creativity, learning, and innovation.

Sandstrom *et al.* [2010] studied the innovation process from the standpoint of both continuous and discontinuous innovation, distinguishing between them and the way they treat ideas in both processes. Consequently, having a two-fold approach stresses the importance of treating discontinuous and continuous ideas in different ways. Therefore, it is not unexpected that the initial stage of the innovation process, where ideas are generated and recognized, has been acknowledged as one of the most important phases of a process that has great impact on the success and costs of innovation [Bjork and Magnusson, 2009].

Van den Ende *et al.* [2015, p. 482] emphasize the importance of ideas in the innovation process, saying that "[i]deas constitute the lifeblood for firms in generating new products or services, new business models, new processes, and bringing about general organizational or strategic change. Ideas are the result of mental activity, and are formulated verbally so that they can be represented, shared, and refined." Therefore, these authors emphasize the importance of defining and formulating an ideation strategy where companies explicitly align their idea generation and selection activities with their innovation strategy. The reason behind this is that innovation strategy is future-directed and it provides the orientation with which new product ideas are judged [van den Ende *et al.*, 2015].

Traditionally, ideation mechanisms are employed to encourage innovations and their success, measuring their contribution in terms of effectiveness and efficiency. On the one hand, front-end effectiveness contemplates the quantity and quality of ideas that are likely to be implemented, and their potential for value generation is included in front-end effectiveness. Efficiency, on the other hand, refers to the speed of idea screening, their translation into concepts, and finally their conversion into project proposals [van den Ende *et al.*, 2015].

The role of idea management within the innovation process itself has been analyzed throughout different innovation models among the literature (for a deeper discussion of different innovation models, refer to Chapter 2 of this book). For instance, Khurana and Rosenthal [1998] adopted a holistic approach and describe a process with four phases: preliminary opportunity identification, product concept definition, product feasibility, and project planning. The model emphasizes the importance of a structured strategy in order to develop new products. Montoya-Weiss and O'Driscoll [2000] argue that new products stem from an idea or, in many cases, the integration of multiple ideas. Therefore, the authors point out that the process of transforming an idea into a robust concept requires definition of the underlying technologies, identification of expected customer benefits, and assessment of the market opportunity. They adopt what is called a funnel model that goes from idea qualification, concept development, concept rating, and finally concept assessment. The proposed model doesn't consider all the elements of the innovation process, but a procedure or practice for dealing with ideas. The model also enhances the use of idea management software as a fundamental tool to manage the process.

Koen *et al.* [2001] present the new concept development (NCD) model entailing opportunity identification, opportunity analysis, idea generation and enrichment, idea selection, and concept definition. This model has a circular approach. "The circular shape is meant to suggest that ideas are expected to flow, circulate, and iterate between and among all the five elements, in any order or combination, and may use one or more elements more than once" [Koen *et al.*, 2001, p. 48]. In the same vein, Husig *et al.* [2003] depict a three-stage process model composed of environmental screening, idea generation, and concept project and business planning. The authors claim that companies who have superior process expertise for front-end activities are more effective in the front-end of the innovation process. Their explanation could be related to the existence of routines established to constantly monitor changes within their ecosystem, in particular their customer base.

Also of interest is the work of Griffiths-Hemans and Grover [2006, p. 37], who in the idea fruition model distinguish three phases: idea creation, idea concretization, and idea commitment. The authors highlight that, "combining the results from intrinsic motivation and access to relevant and diverse knowledge have notable managerial implications. Intrinsic motivation implies that idea originators find the work fun, and innovators want to keep it that way. That they need cooperation from other individuals in different departments entails requiring some ties with them. For the work to continue to be fun, such ties should be the ones the idea originators choose to form."

The well-known work of Hansen and Birkinshaw [2007] depicts the innovation value chain model, in which they identify the following stages in managing ideas: idea generation (in-house idea, cross-pollination, external sourcing), idea conversion (selection, development), and idea diffusion. The authors explain that the first of the three phases in the chain is to generate ideas, and this can happen inside a unit in the firm, across units in a company through an internal collaborative approach, or through collaboration with partners outside the firm.

The second phase is to convert ideas or, more specifically, select ideas for funding, mainly through the screening and ultimately developing them into products or practices, which is the transformation from ideas into initial results. The third phase is to diffuse those products and practices through their dissemination across the organization.

Cooper and Edgett [2008] developed the so-called stage-gate idea to launch the process, a conceptual and operational map for moving new product projects from idea to launch, which was originally proposed by Cooper [1990, 1994]. The stage-gate process moves from point A (idea) to point B (new product). A typical stage-gate process starts with the idea for a new product that is submitted to gate 1, initial screen. This is the starting point of the project and the first decision made is whether to allocate resources to the project or not. If the decision is positive (go), the project advances to the preliminary assessment stage. The criteria used in this stage deal with strategic alignment, project feasibility, magnitude of the opportunity, differential advantage, synergy with the firm's core business and resources, and market attractiveness. This stage is addressed more in detail in Chapter 8.

Also, Westerski *et al.* [2011] distinguish the following phases in the management of ideas: idea generation, idea improvement, idea selection, idea implementation, and idea deployment. The authors define the idea generation phase as the time to reach out to the environment, community, or a specific target and obtain ideas from them. Second, idea improvement is the process of collaborating with other people to improve the ideas that were gathered in the earlier phase. Third, idea selection aims to tap into the numerous ideas received and choose the best ideas. Fourth, idea implementation begins once an idea has been reviewed and selected to be put into practice — the final goal of this stage is to transform these ideas into products and services.

Thus, we have reviewed the different approaches and methods for idea management. We now dig deep into the process of idea generation and shed light on the different techniques and tools that are used for this purpose. We know from the literature that firms that are effective at generating ideas are those that are focused on not only internal sources but also all the potential sources within and outside the boundaries of the firm. New ideas can stem from internal sources (i.e., employees, managers), external sources (i.e., customers, competitors, distributors, and suppliers), and from doing formal research and development [Crawford, 1992].

Shah *et al.* [2003] refer to the importance and variety of different ideation methods that have been developed, which have been classified into two different categories: on the one hand, the intuitive methods; and on the other hand, the logical ones.

The authors argue that intuitive methods "use mechanisms to break what are believed to be mental blocks" while "logical methods involve systematic decomposition and analysis of the problem, relying heavily on technical databases and direct use of science and engineering principles and/or catalogues of solutions or procedures" [Shah *et al.* 2003, p. 112].

Subsequently, Shah *et al.* [2003] classified intuitive methods into germinal, transformational, progressive, organizational, and hybrid categories. Germinal analysis is exemplified by morphological analysis. Ritchey [1998, p. 3] states, "Essentially, general morphological analysis is a method for identifying and investigating the total set of possible relationships or configurations contained in a given problem complex," and adds, "The approach begins by identifying and defining the parameters (or dimensions) of the problem complex to be investigated, and assigning each parameter a range of relevant values or conditions."

Concerning transformational methods, one good example is the well-known SCAMPER system, an idea generation technique that uses action verbs to stimulate creativity. The technique was developed by Eberle [1996] and is described in his book, *Scamper On: Games for Imagination Development*. SCAMPER is an acronym for Substitute, Combine, Adapt, Modify/Magnify, Purpose, Eliminate/Minify, Rearrange/Reverse. The methodology consists of making questions stemming from each of the action verbs and can be used in the development or improvement of products.

Progressive methods are exemplified by C-sketch, also known as collaborative sketching. Collaborative sketching was first proposed at Arizona State University's Design Automation Lab (DAL) and referred as 5-1-4 G. The method was created as an evolution or extension of Method 6-3-5 [Rohrbach, 1969] in which six designers generate three ideas in each of the five subsequent phases. Accordingly, the method 5-1-4 G was also named for the number of designers, ideas, and passes, where the G designates it as a graphically oriented method.

Additionally, organizational methods are illustrated by mind-mapping techniques created by Buzan and Buzan [1996] or the fishbone diagram developed by Ishikawa [1990]. Mind-mapping or idea-mapping has been referred to as visual and non-linear representations of a set of ideas and their relationships. According to Davies [2011, p. 81], "[m]ind-maps comprise a network of connected and related concepts. However, in mind-mapping, any idea can be connected to any other. Free-form, spontaneous thinking is required when creating a mind-map, and the aim of mind-mapping is to find creative associations between ideas. Thus, mind-maps are principally association maps."

The fishbone diagram by Ishikawa is a graphic method that uses a diagram-based approach to find the likely roots of a problem, starting with a thorough analysis of the situation. The fishbone diagram is composed of four steps to use the tool: identifying the problem, working out the major factors involved, identifying possible causes, and finally, analyzing the diagram. The causes are categorized into people (any person involved in the process), methods (policies, procedures, rules, and regulations), machines (including any equipment or tool), materials (any raw material or other element used in the process), measurements (data used to evaluate the quality of the process itself), and environment (conditions in which the process operates).

To wrap up intuitive methods, we refer to hybrid ones, Gordon's [1961] synectics being a good example. Gordon refers to a synectics session as an excursion or a group problem-solving activity wherein persons are stimulated to think creatively under a loosely structured system. The process begins with a problem introduced by a leader who conducts "an excursion" of the group through a series of steps that attempt to determine a solution to the problem [Wilson *et al.*, 1973].

On the other hand, logical methods are classified in two groups: history-based methods and analytical methods. Within the so-called history-based methods, we can refer to the work of Pahl and Beitz [1996]. "In systematic respects, designing is the optimization of given objectives within partly conflicting constraints. Requirements change with time, so that a particular solution can only be optimized for a particular set of circumstances" [Pahl and Beitz, 1996, p. 2].

One of the most well-known methods is TRIZ [Altshuller, 1984]. Mann [2001, p. 123], in his article, *An Introduction to TRIZ: The Theory of Inventive Problem Solving*, refers to TRIZ as "a philosophy, a method, and a collection of problem definition and solving tools and strategies." Mann [2001, p. 124] argues that basically, "TRIZ researchers have encapsulated the principles of good inventive practice and set them into a generic problem-solving framework. The task of problem definers and problem solvers using the large majority of the TRIZ tools thus becomes one in which they have to map their specific problems and solutions to and from this generic framework." According to Ilevbare *et al.* [2013], TRIZ is a set of conceptual solutions to technical problems. The authors stated, "This set of solutions is a collection of the inventive principles, trends of technical evolution and standard solutions as provided by TRIZ. To apply any of these solutions, a specific and factual technical problem is reduced to its essentials and stated in a conceptual format. In its conceptual form, the problem can then be matched with one or more of the conceptual solutions. The identified conceptual solution can afterwards be transformed into a specific, factual solution that answers to the original factual problem" [Ilevbare *et al.*, 2013, p. 31].

Finally, we refer to the analytical methods, such as systematic inventive thinking (SIT), a down-to-earth method applied to creativity, innovation, and problem solving. The bases for SIT's method, which are common with TRIZ principles, stem from the idea that inventive solutions share common patterns.

6.4 ORGANIZATIONAL LEARNING AND KNOWLEDGE MANAGEMENT

Nowadays, knowledge is the cornerstone of building a competitive advantage in organizations. The most distinctive and unique resource available to organizations is embodied knowledge, which facilitates the management of corporate resources. Argote and Ingram [2000, p. 1] argued that, "[b]y embedding knowledge in interactions involving people, organizations can both affect knowledge transfer internally and impede knowledge transfer externally. Thus, knowledge embedded in the interactions of people, tools, and tasks provides a basis for competitive advantage in firms." Kogut and Zander [1992] defend that the firm is a repository of capabilities, and switching to new capabilities is difficult since the social capital of the firm generates inertia in the firm's capabilities. Thus, knowledge-based resources are essential in providing the base for sustainable competitive advantage [McEvily and Chakravarthy, 2002]. Organizational capability to create knowledge is cited by many scholars as the most important source of a firm's sustainable competitive advantage [Junnarkar, 1997; Nonaka *et al.*, 2000].

In particular, novelty and utility are necessary for successful exploitation of new ideas. Anand *et al.* [2007] highlighted the importance of individual expertise and knowledge, which triggers employees to generate novel ideas and innovations. The authors also demonstrated four critical generative elements for knowledge management: socialized agency, differentiated expertise, defensible turf, and organizational support.

Even though knowledge is an important and valuable asset, however, only unique knowledge and not just common knowledge is the source of competitive advantage [Snell *et al.*, 1999]. Lengnick-Hall and Lengnick-Hall [2002] point out that unique knowledge is so specific in content that it is not easily transferred to other organizations because it is difficult to produce innovations based on generic knowledge. Organizations value people with unique KSAs who are new to the firm and to the market and an invaluable resource for innovation [Nonaka and Takeuchi, 1995]. The role of organizational knowledge as a competitive advantage has been emphasized throughout the literature. From the classical perspective, Porter [1980] took a novel approach to knowledge, stating that it integrates organizational knowledge within industry or rivalry perspectives. Therefore, an organizational knowledge-based perspective has appeared in the strategic management approach. This knowledge-based perspective postulates that knowledge is embedded in and carried through multiple entities, including organizational culture and identity, routines, policies, systems, and documents, as well as individual employees [Alavi and Leidner, 2001]. Thus, knowledge-based resources are habitually difficult to copy and socially complex at the same time, which make the knowledge deposits a valuable asset that are the basis for a long-term competitive advantage. Nonetheless, it is not only the fact of knowledge accumulation that constitutes a competitive asset but also the ability to apply this knowledge effectively in the creation of new products, processes, or any kind of value to the market. Being aware of the aforementioned, knowledge management becomes a strategic approach for innovation and competitiveness. Consequently, shedding light on knowledge characteristics and the knowledge creation process would be helpful.

Based on the work of Polanyi [1967] and Nonaka *et al.* [1994], knowledge is categorized into two different dimensions: tacit and explicit. According to Polanyi [1966, as cited in Nonaka *et al.*, 1994, p. 338], "explicit or codified knowledge refers to knowledge that is transmittable in formal, systematic language; whereas tacit knowledge has a personalized quality which makes it hard to formalize and communicate." Nonaka *et al.* [1994] explained that the tacit dimension of knowledge is contained in both cognitive (mental models including mental maps, beliefs, paradigms, and viewpoints) and technical (know-how, crafts, and skills) elements.

According to Nonaka [1994], the knowledge creation process can be structured into four different patterns and involves the use of both tacit and explicit knowledge. Throughout the application of those patterns, existing knowledge can be converted into new knowledge. Bergeron [2003, p. 9] argues that knowledge management "is fundamentally about a systematic approach to managing intellectual assets and other information in a way that provides the company with a competitive advantage. Knowledge Management is a business optimization strategy, and not limited to a particular technology or source of information."

Bergeron [2003] clarifies the different concepts that are related to knowledge management and distinguishes between data, information, metadata, knowledge, and instrumental understanding (Table 6.1).

Assuming that knowledge is generated through the conversion between tacit and explicit knowledge, there are four different modes of knowledge conversion: from tacit knowledge to tacit knowledge; explicit knowledge

Table 6.1. Key Concepts of Knowledge Management.

Concepts	Explanation
Data	"Data represents only numerical quantities or attributes that are derived from experiments, observations, calculations etc." [Bergeron, 2003, p. 10].
Information	"Information consists of data associated to particular explanations, interpretations, or other textual material concerning a particular object/process" [Bergeron, 2003, p. 10].
Metadata	"Metadata represents additional information regarding the context in which main information is used" [Bergeron, 2003, p. 10].
Knowledge	"Knowledge is defined as information, which is organized and synthesized in order to foster comprehension, awareness, and understanding. Knowledge combines both metadata and awareness of the context suitable for applying metadata" [Bergeron, 2003, p. 10].
Instrumental understanding	"Instrumental understanding is the clear and complete perception of the nature and significance of an issue. It is the internal capability to gain experience by relating specific knowledge to broader themes. Other researchers refer to this concept as either sense-making or wisdom" [Bergeron, 2003, p. 10].

Source: Bergeron [2003].

Table 6.2. Four Types of Knowledge Conversion.

Conversion mode	Explanation
Socialization: Tacit knowledge to tacit knowledge	Socialization is the process of transforming tacit knowledge into more tacit knowledge. A good example of this is apprenticeship, where one individual acquires new knowledge throughout an experiential activity. It is even possible to socialize knowledge without conversation or the use of language. "Without some form of shared experience, it is extremely difficult for people to share each other's thinking processes" [Nonaka *et al.*, 1994, p. 340].
Combination: Explicit knowledge to explicit knowledge	Combination is the process of transforming explicit knowledge to explicit knowledge. "Individuals exchange and combine knowledge through such exchange mechanisms as meetings and telephone conversations" [Nonaka *et al.*, 1994, p. 340]. The point is combining already existing knowledge through social processes.
Externalization: Tacit knowledge to explicit knowledge	Externalization is the process of transforming tacit knowledge into explicit knowledge. "This interaction involves two different operations. One is the conversion of tacit knowledge into explicit knowledge, which we call externalization" [Nonaka *et al.*, 1994, p. 340].
Internalization: Explicit knowledge to tacit knowledge	Internalization is the process of transforming explicit knowledge into tacit knowledge. "The other is the conversion of explicit knowledge into tacit knowledge, which we call "internalization" and which is similar to the traditional notion of learning" [Nonaka *et al.*, 1994, p. 340].

Source: Nonaka *et al.* [1994].

to explicit knowledge; tacit knowledge to explicit knowledge; and explicit knowledge to tacit knowledge [Nonaka *et al.*, 1994, p. 340]. In Table 6.2, we find explanations of the different conversion modes.

Considering the aforementioned, the transfer of knowledge within the organization is the cornerstone of the competitiveness of the firm. Argote and Ingram [2000, p. 151] defined knowledge transfer in organizations as "the process through which one unit (e.g., group, department, or division) is affected by the experience of another." The authors referred to knowledge transfer, recognizing that the basis applied at the individual and organizational levels are the same. However, they point out that "knowledge transfer in organizations involves transfer at the individual level; the problem of knowledge transfer in organizations transcends the individual level to include transfer at higher levels of analysis, such as the group, product line, department, or division" [Argote and Ingram, 2000, p. 151]. Accordingly, knowledge creation occurs continuously within the organization in a process where individuals, teams, and groups within the firm and between groups outside the boundaries of the firm share all kinds of knowledge [Bloodgood and Salisbury, 2001].

Specifically, to properly manage knowledge at the organizational level, the business has to be aware of the knowledge deposits or reservoirs in the firm. Argote and McGrath [1993] state that knowledge is embedded in an organization's members (people involved in the organization), tools (the technological component of the firm), and tasks (goals, intentions, and purposes), as well as the combination of these elements.

Knowledge transfer in organizations can be addressed under two approaches. The first of these determines the ability to generate, stock, share, and use the documented knowledge in the firm, which stresses the importance of codifying and storing knowledge. Authors like Swan *et al.* [2000] determined the potential impact of IT on knowledge management for innovation and the relationship between knowledge management and the existing social networks and communities within organizations. This approach can be referred as a system strategy [Choi and Lee, 2002] since the organization identifies all the knowledge within the organization as well as all the elements that can act as knowledge reservoirs in the firm, and finally the processes of knowledge transfer between the different units in the firm.

There is another approach that highlights the importance of knowledge sharing, taking into account personal interactions. This approach stresses the importance of dialogue through social networks, teams, and groups [Swan *et al.*, 2000], personal interaction [Hansen *et al.*, 1999], and the identification of knowledge that is shared informally [Jordan and Jones, 1997]. This approach can be referred to as the human strategy [Choi and Lee, 2002], since it stresses the interaction between people, not the system itself, as the basis for knowledge sharing and transfer.

Bloodgood and Salisbury [2001] distinguish three ways in which the organization manages its knowledge: knowledge creation, knowledge transfer, and knowledge protection. In order to differentiate the three different strategies, the authors argue, "Speaking in broad terms, organizations that use a strategy of knowledge creation focus on creativity, experimentation, and, to a significant extent, creating a shared understanding within the creating group to construct new knowledge that can be used to develop new products and services" [Bloodgood and Salisbury, 2001 p. 58]. Second, the authors state that organizations using a strategy of knowledge transfer prioritize the dissemination of knowledge throughout the firm to use it rapidly. Finally, the knowledge protection strategy emphasizes the protection of knowledge in its original form.

Holsapple and Joshi [2002, p. 49] address knowledge management by distinguishing three steps: characterizing knowledge resources (knowledge reservoirs represented in tacit and explicit knowledge) that need to be managed (processing knowledge resources); identifying and explaining activities involved in manipulating these knowledge resources; and recognizing factors that influence the conduct of knowledge management (knowledge resource components, activities, and factors influencing the knowledge management process).

Holsapple and Joshi [1999] also offer a framework for knowledge management in which they portray acquisition (identifying knowledge outside the boundaries of the firm and transforming it for the purpose of the firm throughout the internalization); selection (identifying knowledge inside the boundaries of the firm being used within the firm); internalizing (growth or modification of the existing knowledge by the assimilation of new knowledge); and, finally, using (generation of new knowledge or its externalization outside the organization) knowledge.

Once we have studied how organizations identify, capture, and manage new knowledge, it is also important to explore the processes of organizational learning within the business. Organizational learning also has a clear strategic approach. According to Crossan *et al.* [1999, p. 522], organizational learning can be conceived of as "a principal means of achieving the strategic renewal of an enterprise." The authors also argued that "strategic renewal places additional demands on a theory of organizational learning. Renewal requires that organizations explore and learn new ways while concurrently exploiting what they have already learned." Organizational learning goes beyond knowledge management, since it includes processes such as control and intelligence or exploitation of

Table 6.3. Kinds of Organizational Learning.

Processes	Construct	Explanation
Knowledge acquisition	Congenital learning	"Organization's congenital knowledge is a combination of the knowledge inherited at its conception and the additional knowledge acquired prior to its birth" [Huber, 1991, p. 91].
	Experiential learning	"(1) Organizational experiments, (2) organizational self-appraisal, (3) experimenting organizations, (4) unintentional or unsystematic learning, and (5) experience-based learning curves" [Huber, 1991, p. 91].
	Vicarious learning	Vicarious learning is also known as acquiring second-hand experience. "Organizations commonly attempt to learn about the strategies, administrative practices, and especially technologies of other organizations" [Huber, 1991, p. 96].
	Grafting	"Organizations frequently increase their store of knowledge by acquiring and grafting on new members who possess knowledge not previously available within the organization" [Huber, 1991, p. 97].
	Searching and noticing	"Organizational information acquisition through search can be viewed as occurring in three forms: (1) scanning, (2) focused search, and (3) performance monitoring" [Huber, 1991, p. 97].
Information distribution		"Information distribution is a determinant of both the occurrence and breadth of organizational learning. With regard to occurrence of organizational learning, consider that organizational components commonly develop 'new' information by piecing together items of information that they obtain from other organizational units, as when a shipping department learns that a shortage problem exists by comparing information from the warehouse with information from the sales department" [Huber, 1991, p. 100].

(Continued)

Table 6.3. *(Continued)*

Processes	Construct	Explanation
Information interpretation	Cognitive maps and framing	"The facts that a person's prior cognitive map (or belief structure or mental representation or frame of reference) will shape his or her interpretation of information, and that these cognitive maps vary across organizational units having different responsibilities, are well established" [Huber, 1991, p. 102]
	Media richness	"Media richness is a determinant of the extent to which information is given common meaning by the sender and receiver of a message" [Huber, 1991, p. 103].
	Information overload	"Overload detracts from effective interpretation" [Huber, 1991, p. 104].
	Unlearning	"… unlearning can lead to either a decrease, or an increase, in the range of potential behaviors. It follows then, from the earlier definition of learning, that unlearning is conceptually subsumable under learning" [Huber, 1991, p. 104]
Organizational memory	Storing and retrieving information	"Organizations store a great deal of "hard" information on a routine basis, sometimes for operating reasons and sometimes to satisfy the reporting requirements of other units or organizations" [Huber, 1991, p. 105].
	Computer-based organizational memory	"Information concerning the times necessary to complete fabrication of certain products, to receive shipments of ordered materials, to recruit or train various types of employees, or to deliver certain types of services are more and more frequently resident in computers as transactions artifacts, either those created and transmitted internally using the organization's electronic mail, electronic bulletin board, or electronic blackboard systems, or those exchanged electronically across the organization's boundaries (e.g., letters, billings, and contracts)" [Huber, 1991, p. 106].

Source: Huber [1991].

knowledge and technology [Templeton *et al.*, 2002]. According to Huber [1991, p. 89], organizational learning is defined as a process of outlining the learning itself. "An entity learns if, through its processing of information, the range of its potential behaviors is changed [...] an organization learns if any of its units acquires knowledge that it recognizes as potentially useful to the organization." Likewise, the author claims that the breadth of organizational learning depends on more of the organization's components used to obtain this knowledge and recognizes it as potentially useful when more and more varied interpretations are developed because such development changes the range of potential behaviors and when more organizational units develop uniform comprehensions of the various interpretations [Huber, 1991, p. 90].

Huber [1991] depicted an organizational learning model in four phases or stages as follows:

(1) Knowledge acquisition as the process by which knowledge is obtained. It helps in distinguishing congenital learning, experiential learning, vicarious learning, grafting, and searching and noticing.
(2) Information distribution as the process by which information from different sources is shared and thereby leads to new information or understanding.
(3) Information interpretation as the process by which distributed information is given one or more commonly understood interpretations. It helps in distinguishing cognitive maps and framing, media richness, information overload, and unlearning.
(4) Organizational memory as the means by which knowledge is stored for future use. It helps in distinguishing, storing, and retrieving information and computer-based organizational memory.

In Table 6.3, you can find a description of each of the aforesaid kinds of organizational learning.

REFERENCES

Alavi, M., and Leidner, D. E. [2001]. Knowledge management and knowledge management systems: Conceptual foundations and research issues. *MIS Quarterly*, pp. 107–136.

Altshuller, G. S. [1984]. *Creativity as an Exact Science: The Theory of the Solution of Inventive Problems* (Gordon and Breach, UK).

Amabile, T. M. [1996]. *Creativity in Context: Update to the Social Psychology of Creativity*, (Hachette, UK).

Amabile, T. M. [1997]. Motivating creativity in organizations: On doing what you love and loving what you do, *California Management Review*, 40(1), pp. 39–58.

Amabile, T. M. [1998]. How to kill creativity, *Harvard Business Review*, 76(5), pp. 76–87.

Anand, N., Gardner, H. K., and Morris, T. [2007]. Knowledge-based innovation: Emergence and embedding of new practice areas in management consulting firms, *Academy of Management Journal*, 50(2), pp. 406–428.

Anderson, N., De Dreu, C. K., and Nijstad, B. A. [2004]. The routinization of innovation research: A constructively critical review of the state-of-the-science, *Journal of Organizational Behavior*, 25(2), pp. 147–173.

Argote, L., and McGrath, J. E. [1993]. Group processes in organizations: Continuity and change, *International Review of Industrial and Organizational Psychology*, 8(1993), pp. 333–389.

Argote, L., and Ingram, P. [2000]. Knowledge transfer: A basis for competitive advantage in firms, *Organizational Behavior and Human Decision Processes*, 82(1), pp. 150–169.

Barney, J. B. [1995]. Looking inside for competitive advantage, *Academy of Management Perspectives*, 9(4), pp. 49–61.

Barron, F. [1955]. The disposition toward originality. *The Journal of Abnormal and Social Psychology*, 51(3), p. 478.

El Bassiti, L., and Ajhoun, R. [2013]. Toward an innovation management framework: A life-cycle model with an idea management focus, *International Journal of Innovation, Management and Technology*, 4(6), pp. 551–559.

Becker, G. S. [1975]. *Human Capital: A Theoretical and Empirical Analysis, with Special Reference to Education*, 2nd edn. (NBER, USA).

Bergeron, B. [2003]. *Essentials of Knowledge Management,* Vol. 28 (John Wiley & Sons, USA).

Björk, J., and Magnusson, M. [2009]. Where do good innovation ideas come from? Exploring the influence of network connectivity on innovation idea quality, *Journal of Product Innovation Management*, *26*(6), pp. 662–670.

Bloodgood, J. M., and Salisbury, W. D. [2001]. Understanding the influence of organizational change strategies on information technology and knowledge management strategies, *Decision Support Systems*, 31(1), pp. 55–69.

Boeddrich, H. J. [2004]. Ideas in the workplace: A new approach towards organizing the fuzzy front end of the innovation process, *Creativity and Innovation Management*, 13(4), pp. 274–285.

Bontis, N., Dragonetti, N. C., Jacobsen, K., and Roos, G. [1999]. The knowledge toolbox: A review of the tools available to measure and manage intangible resources, *European Management Journal*, 17(4), pp. 391–402.

Booz, Allen, and Hamilton. [1982]. *New Product Management for the 1980's*, (Booz, Allen and Hamilton, New York, USA).

Björk, J., and Magnusson, M. [2009]. Where do good innovation ideas come from? Exploring the influence of network connectivity on innovation idea quality, *Journal of Product Innovation Management,* 26(6), pp. 662–670.

Bornay-Barrachina, M., De la Rosa-Navarro, D., López-Cabrales, A., and Valle-Cabrera, R. [2012]. Employment relationships and firm innovation: The double role of human capital, *British Journal of Management*, 23(2), pp. 223–240.

Buzan, T., and Buzan, B. [1996]. *How to Use Radiant Thinking to Maximize Your Brain's Untapped Potential* (Dutton, New York, USA).

Choi, B., and Lee, H. [2002]. Knowledge management strategy and its link to knowledge creation process, *Expert Systems with Applications*, 23(3), pp. 173–187.

Cooper, R. G. [1990]. Stage-gate systems: A new tool for managing new products, *Business Horizons,* 33(3), pp. 44–54.

Cooper, R. G. [1994]. Perspective third-generation new product processes, *Journal of Product Innovation Management*, 11(1), pp. 3–14.

Cooper, R. G., and Edgett, S. J. [2008]. Maximizing productivity in product innovation, *Research-Technology Management*, 51(2), pp. 47–58.

Crawford, C. [1992]. The hidden costs of accelerated product development, *Journal of Product Innovation Management*, 9(3), pp. 188–199.

Crossan, M. M., Lane, H. W., and White, R. E. [1999]. An organizational learning framework: From intuition to institution, *Academy of Management Review*, 24(3), pp. 522–537.

Dakhli, M., and De Clercq, D. [2004]. Human capital, social capital, and innovation: A multi-country study, *Entrepreneurship & Regional Development*, 16(2), pp. 107–128.

Davies, M. [2011]. Concept mapping, mind mapping and argument mapping: What are the differences and do they matter? *Higher Education*, 62(3), pp. 279–301.

Eberle, B. [1996]. *Scamper On: Games for Imagination Development* (Prufrock Press Inc, USA).

Farr, J. L., and West, M. A. (eds.), [1990]. *Innovation and Creativity at Work: Psychological and Organizational Strategies* (Wiley, USA).

Ford, C. M. [1996]. A theory of individual creative action in multiple social domains, *Academy of Management Review*, 21(4), pp. 1112–1142.

Francis, D., and Bessant, J. [2005]. Targeting innovation and implications for capability development, *Technovation*, 25(3), pp. 171–183.

Galbraith, J. R. [1982]. Designing the innovating organization, *Organizational Dynamics*, 10(3), pp. 5–25.

Gilson, L. L., and Shalley, C. E. [2004]. A little creativity goes a long way: An examination of teams' engagement in creative processes, *Journal of Management*, 30(4), pp. 453–470.

Gordon, W. J. [1961]. *Synectics: The Development of Creative Capacity* (Harper, Oxford, UK)

Grant, R. M. [1995]. *Contemporary Strategy Analysis: Concepts, Techniques, Applications* (Blackwell, USA).

Green, S. G., Bean, A. S., and Snavely, B. K. [1983]. Idea management in R&D as a human information processing analog, *Human Systems Management*, 4(2), pp. 98–112.

Griffiths-Hemans, J., and Grover, R. [2006]. Setting the stage for creative new products: Investigating the idea fruition process, *Journal of the Academy of Marketing Science*, 34(1), pp. 27–39.

Hansen, M. T., and Birkinshaw, J. [2007]. The innovation value chain, *Harvard Business Review*, 85(6), pp. 121–130.

Hansen, M. T., Nohria, N., and Tierney, T. [1999]. What's your strategy for managing knowledge? *Harvard Business Review*, 77(2), pp. 106–116.

Holsapple, C.W., and Joshi, K. D. [1999]. *Knowledge Management Handbook*, Liebowitz, J. (ed.), Chapter 7, *Knowledge Selection: Concepts, Issues and Technologies* (CRC Press, Boca Raton) pp. 7.1–7.17.

Holsapple, C. W., and Joshi, K. D. [2002]. Knowledge management: A threefold framework, *The Information Society*, 18(1), pp. 47–64.

Huber, G. P. [1991]. Organizational learning: The contributing processes and the literatures, *Organization Science*, 2(1), pp. 88–115.

Ilevbare, I. M., Probert, D., and Phaal, R. [2013]. A review of TRIZ, and its benefits and challenges in practice, *Technovation*, 33(2–3), pp. 30–37.

Ishikawa, Kaoru [1990]. *Introduction to Quality Control* (3A Corporation, Tokyo, Japan).

Jordan, J., and Jones, P. [1997]. Assessing your company's knowledge management style, *Long Range Planning*, 30(3), pp. 392–398.

Kamoche, K. [1996]. Strategic human resource management within a resource-capability view of the firm, *Journal of Management Studies*, 33(2), pp. 213–233.

Kimberly, J. R., and Evanisko, M. J. [1981]. Organizational innovation: The influence of individual, organizational, and contextual factors on hospital adoption of technological and administrative innovations, *Academy of Management Journal*, 24(4), pp. 689–713.

Khurana, A., and Rosenthal, S. R. [1998]. Towards holistic "front ends" in new product development, *Journal of Product Innovation Management: An International Publication of the Product Development & Management Association*, 15(1), pp. 57–74.

Koen, P., Ajamian, G., Burkart, R., Clamen, A., Davidson, J., D'Amore, R., and Karol, R. [2001]. Providing clarity and a common language to the "fuzzy front end", *Research-Technology Management*, 44(2), pp. 46–55.

Koen, P. A., Ajamian, G. M., Boyce, S., Clamen, A., Fisher, E., Fountoulakis, S., and Seibert, R. [2002]. *Fuzzy Front End: Effective Methods, Tools, and Techniques. The PDMA Tool Book 1 for New Product Development* (PDMA, USA).

Kogut, B., and Zander, U. [1992]. Knowledge of the firm, combinative capabilities, and the replication of technology, *Organization Science*, 3(3), pp. 383–397.

Laursen, K. [2002]. The importance of sectoral differences in the application of complementary HRM practices for innovation performance, *International Journal of the Economics of Business*, 9(1), pp. 139–156.

Lengnick-Hall, M., and Lengnick-Hall, C. [2002]. *Human Resource Management in the Knowledge Economy: New Challenges, New Roles, New Capabilities* (Berrett-Koehler Publishers, USA).

Lepak, D. P., Takeuchi, R., and Snell, S. A. [2003]. Employment flexibility and firm performance: Examining the interaction effects of employment mode, environmental dynamism, and technological intensity, *Journal of Management*, 29(5), pp. 681–703.

Lopez-Cabrales, A., Valle, R., and Herrero, I. [2006]. The contribution of core employees to organizational capabilities and efficiency, *Human Resource Management*, 45(1), pp. 81–109.

McEvily, S. K., and Chakravarthy, B. [2002]. The persistence of knowledge-based advantage: An empirical test for product performance and technological knowledge, *Strategic Management Journal*, 23(4), pp. 285–305.

Mann, D. [2001]. An introduction to TRIZ: The theory of inventive problem solving, *Creativity and Innovation Management*, 10(2), pp. 123–125.

Marqués, D. P., Simón, F. J. G., and Carañana, C. D. [2006]. The effect of innovation on intellectual capital: An empirical evaluation in the biotechnology and telecommunications industries, *International Journal of Innovation Management*, 10(01), pp. 89–112.

Martins, E. C., and Terblanche, F. [2003]. Building organizational culture that stimulates creativity and innovation, *European Journal of Innovation Management*, 6(1), pp. 64–74.

Mikelsone, E., and Liela, E. [2015]. Discussion on the terms of idea management and idea management systems, *Regional Formation and Development Studies*, 17(3), pp. 97–111.

Montoya-Weiss, M. M., and O"Driscoll, T. M. [2000]. From experience: Applying performance support technology in the fuzzy front end, *Journal of Product Innovation Management*, 17(2), pp. 143–161.

Mumford, M. D., Scott, G. M., Gaddis, B., and Strange, J. M. [2002]. Leading creative people: Orchestrating expertise and relationships, *The leadership Quarterly*, 13(6), pp. 705–750.

Nahapiet, J., and Ghoshal, S. [1998]. Social capital, intellectual capital, and the organizational advantage, *Academy of Management Review*, 23(2), pp. 242–266.

Nonaka, I., Byosiere, P., Borucki, C. C., and Konno, N. [1994]. Organizational knowledge creation theory: A first comprehensive test, *International Business Review*, 3(4), pp. 337–351.

Nonaka, I. [1994]. A dynamic theory of organizational knowledge creation, *Organization Science*, 5(1), pp. 14–37.

Nonaka, I., and Takeuchi, H. [1995]. *The Knowledge-Creating Company: How Japanese Companies Create the Dynamics of Innovation* (Oxford University Press, UK).

Nonaka, I., Toyama, R., and Konno, N. [2000]. SECI, ba and leadership: A unified model of dynamic knowledge creation, *Long Range Planning*, 33, pp. 5–34

Nyström, H. [1990]. *Technological and Market Innovation: Strategies for Product and Company Development*, 1st edn. (Wiley, Chichester).

Oldham, G. R., and Cummings, A. [1996]. Employee creativity: Personal and contextual factors at work, *Academy of Management Journal*, 39(3), pp. 607–634.

Pahl, G., and Beitz, W. [1996]. *Engineering a Systematic Approach* (Springer, Germany).

Paulus, P. B., Dzindolet, M., and Kohn, N. W. [2012]. Handbook of organizational creativity, Mumford, D. (eds.), Chapter, 14 *Collaborative Creativity — Group Creativity and Team Innovation* (Academic Press, Cambridge, Massachusetts) pp. 327–357.

Polanyi, M. [1966]. The logic of tacit inference, *Philosophy*, 41(155), pp. 1–18.

Polanyi, M. [1967]. *The Tacit Dimension* (Anchor Books, Garden City, NY).

Porter, M. E. [1980]. Industry structure and competitive strategy: Keys to profitability, *Financial Analysts Journal*, 36(4), pp. 30–41.

Ritchey, T. [1998]. General morphological analysis, *Proceedings of the 16th Euro Conference on Operational Analysis*, Swedish Morphological Society.

Rowbotham, L., and Bohlin, N. [1996]. Structured idea management as a value-adding process, *Prism*, pp. 79–94.

Shah, J. J., Smith, S. M., and Vargas-Hernandez, N. [2003]. Metrics for measuring ideation effectiveness, *Design Studies*, 24(2), pp. 111–134.

Shalley, C. E., and Gilson, L. L. [2004]. What leaders need to know: A review of social and contextual factors that can foster or hinder creativity, *The Leadership Quarterly*, 15(1), pp. 33–53.

Sandstrom, C., and Bjork, J. [2010]. Idea management systems for a changing innovation landscape, *International Journal of Product Development*, 11(3–4), pp. 310–324.

Schultz T. W. [1981]. *Investing in People. The Economics of Population Quality* (University of California Press, Berkeley, USA).

Schroeder, R. G., Van de Ven., A. H., Scudder, G. D. and Polley, D. [2000]. *Research on the Management of Innovation,* Van de Ven A. H. *et al.* (eds.), Chapter 4, *The Development of Innovative Ideas* (Oxford University Press, UK), pp. 107–134

Smith, K. G., Collins, C. J., and Clark, K. D. [2005]. Existing knowledge, knowledge creation capability, and the rate of new product introduction in high-technology firms, *Academy of Management Journal*, 48(2), pp. 346–357.

Snell, S. A., Lepak, D. P., and Youndt, M. A. [1999]. Managing the architecture of intellectual capital: Implications for strategic human resource management,

Research in Personnel and Human Resources Management, 4(1), pp. 175–193.

Subramaniam, M., and Youndt, M. A. [2005]. The influence of intellectual capital on the types of innovative capabilities, *Academy of Management Journal*, 48(3), pp. 450–463.

Swan, J., Newell, S., and Robertson, M. (2000, January). Limits of IT-driven knowledge management initiatives for interactive innovation processes: towards a community-based approach, *Proceedings of the 33rd Annual Hawaii International Conference on System Sciences, IEEE*, 11 pages.

Teece, D. J. [2010]. Business models, business strategy and innovation, *Long Range Planning*, 43(2–3), pp. 172–194.

Templeton, G. F., Lewis, B. R., and Snyder, C. A. [2002]. Development of a measure for the organizational learning construct, *Journal of Management Information Systems*, 19(2), pp. 175–218.

Thornhill, S. [2006]. Knowledge, innovation and firm performance in high-and low-technology regimes, *Journal of Business Venturing*, 21(5), pp. 687–703.

Tidd, J. E., Bessant, J., and Pavitt, K. [2009]. *Managing Innovation: Integrating Technological, Market and Organizational Change* (Wiley, USA).

Tierney, P., Farmer, S. M., and Graen, G. B. [1999]. An examination of leadership and employee creativity: The relevance of traits and relationships, *Personnel Psychology*, 52(3), pp. 591–620.

Ulrich, D., and Lake, D. [1991]. Organizational capability: Creating competitive advantage, Academy *of Management Perspectives*, 5(1), pp. 77–92.

Van den Ende, J., Frederiksen, L., and Prencipe, A. [2015]. The front end of innovation: Organizing search for ideas, *Journal of Product Innovation Management*, 32(4), pp. 482–487.

Van de Ven, A. H. [1986]. Central problems in the management of innovation, *Management Science*, 32(5), pp. 590–607.

Westerski, A., Iglesias, C. A., and Nagle, T. [2011]. The road from community ideas to organisational innovation: A life cycle survey of idea management systems, *International Journal of Web Based Communities*, 7 [4], pp. 493–506.

Wilson, S. H., Greer, J. F., and Johnson, R. M. [1973]. Synectics, a creative problem-solving technique for the gifted, *Gifted Child Quarterly*, 17(4), pp. 260–267.

Woodman, R. W., Sawyer, J. E., and Griffin, R. W. [1993]. Toward a theory of organizational creativity, *Academy of Management Review*, 18(2), pp. 293–321.

CHAPTER 7

INNOVATION IN PROCESSES

7.1 DEFINING BUSINESS PROCESS

The term *process* is a multifaceted concept that is used with different connotations and within different environments, and until now, we have not offered a clear definition. Accordingly, Palmberg [2009] argues that the term "process" has been defined in multiple ways and that it's not easy to find a single definition that everyone can agree on. We gathered the many approaches or factors that can be found in process characterization among the literature, depicting six components: input and output orientation, the existence of interrelated activities, intra-functional or cross-functional approaches, purpose or value to customer, repeatability, and the involvement and use of resources.

To shed light on what the term *process* really means, we refer to the scientific literature. For instance, Davenport [1993, p. 6] defines process as "a structured, measured set of activities designed to produce a specified output for a particular customer or market. It implies a strong emphasis on how work is done within an organization, in contrast to a product focus's emphasis on what" and adds that "a process is thus a specific ordering of work activities across time and place, with a beginning, an end, and clearly identified inputs and outputs: a structure for action" encompassing much of the components mentioned earlier. Furthermore, Llewellyn and Armistead [2000, p. 225] define a business process as "a series of interrelated activities that cross functional boundaries with individual inputs and outputs" stressing the importance of three of the components of the definition, the interrelation of elements, the cross-functional character, and the input–output approach. Also, Hammer and Champy [1993] define business processes as a collection of activities that takes one or more kinds of input and creates an output that is of value to the customer, adding in this case the customer or value orientation as part of the definition.

The input–output element is a common component of the definition that we have found in the literature that establishes the reference to the involvement of a series of resources managed in a particular way to obtain a series of outputs or results that can be traced and measured. This view is consistent with the importance of the measurement of efficiency aiming at a continuous improvement of the process. Correspondingly, Palmberg [2009, p. 207] proposes a process definition as "a horizontal sequence of activities that transforms an input (need) to an output (result) to meet the needs of customers or stakeholders," stressing again the importance of customers' needs. Again, Palmberg [2009, p. 207] emphasizes the nature of the process as a set of interrelated activities depicting its own nature and composition, where there exists a hierarchy within the process structure: process, subprocess, activity, and tasks. Also, Talwar [1993, p. 26] defines process as "a sequence of predefined activities executed to achieve a prespecified type or range of outcomes," adding that, ideally, the result should satisfy the stakeholders of the process, such as customers, employees, and management. Likewise, Zairi [1997, p. 64] defined a process as "an approach for converting inputs into outputs. It is the way in which all the resources of an organization are used in a reliable, repeatable, and consistent way to achieve its goals." Another view of the term defines process as "a set of attributes and principled flow of steps in order to achieve a task. In general, process helps in governing the operations of an organization such that it can produce valuable outputs" [Anand *et al.*, 2013, p. 2].

Focusing on the structure or process framework, we mention the work of Dumas *et al.* [2013, p. 4], who argue that a business process encompasses a number of events and activities. The authors reason that the difference between events and activities lies in the existence of a specific duration allocated to it; while the event has no duration, the activity involves a period of time to complete it. They also specify that the activity can be split into different tasks or single work units and that it involves a number of actors (human actors, organizations, or software systems acting on behalf of human actors or organizations), physical objects (equipment, materials, products, and paper documents), and immaterial objects (electronic documents and electronic records). Finally, the process also leads to one or several outcomes and its delivery to actors that are involved in the process.

Process conceptualization corresponds to one of two organizational perspectives or orientations: the production theory and system-oriented organization theory. On the one hand, following the production perspective, it views "organizations as tools for goal attainment; this approach assesses effectiveness in terms of attainment of clearly defined objectives and production of specific outputs" [Harrison, 2004, p. 41]. Under this perspective, a process is distinctly identified by the input–output and start–finish approach. Conversely, a system-oriented theory focuses on the interrelation of the elements of the organization and between the organization and its environment as a system. On the one hand, focusing on the production function, we can cite Utterback and Abernathy [1975, p. 641] who define the production process as "the system of process equipment, work force, task specifications, material inputs, work and information flows, etc. that are employed to produce a product or service." On the other hand, Teece *et al.* [1997, p. 518] define managerial and organizational processes as "the way things are done in the firm, or what might be referred to as its routines or patterns of current practice and learning." These authors link the firm's processes and positions collectively to its competences and capabilities, constituting the base of competitive advantages.

Continuing with the categorization of processes in the firm, Talwar [1993] emphasizes the importance of differentiating between business functions and business processes. The authors argue that "organizations can be developed in a function-oriented or process-oriented manner. A function-oriented organization divides similar tasks into functions (e.g., research and development, purchasing), which then perform these tasks. A process-oriented organization forms key processes, which take place within an organization or between different organizational units" [Schallmo *et al.*, 2018, p. 15]. Similarly, Davenport [1993] discriminates between operational and management processes in manufacturing firms. Primary, operational processes include product development, customer acquisition, customer requirements identification, manufacturing, integrated logistics, order management, and post-sales services. Then, management processes entail performance monitoring, information management, asset management, human resource management, planning, and resource allocation. Similarly, Llewellyn and Armistead [2000, p. 225] categorize business processes "as operational or supporting. Operational processes are associated with the way organizations develop strategies, invent products and services, market and sell these, manage production and delivery of products or services, and bill customers. Support processes include the provision of HRM activities, information systems infrastructure, finance, and asset management." Also, Palmberg [2009] differentiates between strategic management processes, operational delivery processes, and supportive administrative processes. Business processes can be divided into operational processes, activities involving a firm's value chain, and management process (information processing, control, coordination, and communication) [Mooney *et al.*, 1996]. In particular, Brown [2008] distinguishes between six different types of processes: production of goods and services, distribution and logistics, marketing and sales, information and communication systems, administration and management, and product and business process development.

7.2 BUSINESS PROCESS INNOVATION

7.2.1 *Understanding the Benefits of Process Innovation*

As we stated earlier in Chapter 4 of this book, the strategic approach to firm competitiveness is two-fold. On the one hand, keeping the pace in launching new products and services to market must be seen as a guarantee to maintain differentiation positioning with competitors. On the other hand, the firm has to keep an eye on efficiency, so that the overall cost structure is efficient and flexible and the firm will be able to compete in evermore demanding markets. Although the question has frequently been presented as a trade-off, the reality is that businesses have to accomplish both efficiency and innovation, since rapid market and technology evolution pushes firms to continuous adaptation and improvement in business products and processes. Changes in business processes have traditionally been related to cost optimization, quality improvement, time reduction, and increased revenue [Morris and Brandon, 1993].

Even though process innovation seems to be less visible and important than product innovation, radical and incremental process innovations have shown they can profoundly change the patterns of industrial progress. Thus, Kline and Rosenberg [2010, p. 282] stated the importance and relevance of innovation in processes and incremental changes for the business, arguing that "a large part of the technological innovation that is carried out in industrial societies takes the form of very small changes, such as minor modifications in the design of a machine" and add, "in innovation after innovation it is the subsequent improvement process within the framework of an initial innovation, that transforms a mere novelty to a device of great economic significance. There are many instances in which the learning associated with cumulative production of a given item reduced costs by a factor of two or three, including airline costs per passenger-seat-mile, automobiles, and industrial chemicals" [Kline and Rosenberg, 2010, p. 284]. Also, the authors point out that in industries concerned with production of materials for sale to end producers of goods (steel, rubber, semiconductors), "nearly the only technical innovations that bear on profit are process innovations" [Kline and Rosenberg, 2010, p. 292].

In particular, production processes are the basis for efficiency and productivity gains and new models of generation and delivery of goods. Processes are everywhere; every organization, be it a governmental body, a non-profit organization, or an enterprise, has to manage a number of processes [Dumas *et al.*, 2013]. The results of process development drive a variety of effects, such as cost reduction and production increases [Lim *et al.*, 2006], product quality improvement [Pisano, 1997], faster product development through the implementation of new product development (NPD) processes and reduced time to market [Gold, 1987], and better strategic resource allocations [Yoshida, 1998], which is an important asset for overall firm competitiveness. Also, according to the work of Teece *et al.* [1997, p. 518], we stress the relevance of business processes, since "the competitive advantage of firms lies with its managerial and organizational processes, shaped by its asset position, and the paths available to it," where they recognize processes as a cornerstone for the competitive advantage of the firm.

It is not surprising that improving business processes has gained the attention of scholars and practitioners. For instance, Utterback and

Abernathy [1975] discuss the changing nature of production processes that, according to them, evolve over time, gaining productivity through the use of capital, greater division of work, product standardization, and streamlining the flow of materials, driving a cumulative effect that changes the nature of the process and that is accompanied by changes in internal organizational structure, suppliers, and technology-based capital goods. Thus, the authors underline the importance and systemic character of production processes, which influence not only the production function also but the overall business organization and technological base. Accordingly, though product and process innovations seem to be two separate elements in business development, we also have to assume that process innovation in industry is closely related to product innovations. In this regard, we cite again the work of Utterback and Abernathy [1975] and Utterback [1994], who presented a dynamic model of process and product innovation relating the rate of innovation in the industry to the different stages of development. In this regard, Utterback and Abernathy [1975] refer to the process of market development as uncoordinated, segmental, and systemic.

First, the authors argue that the uncoordinated aspect is explained by arguing that in early stages, market and continuous redefinition lead to competitive improvements and high rates of both product and process changes. "Process itself is composed largely of unstandardized and manual operations, or operations that rely upon general purpose equipment. During this state, the process is fluid, with loose and unsettled relationships between process elements" [Utterback and Abernathy, 1975, p. 641].

Second, segmental refers to the stage when an industry matures, which evolves into price competition and mechanistic and elaborate production systems (operating controls). Consequently, production systems lead to automation and process controls and some highly automated subprocesses, while others are still manually implemented. "As a result, production processes in this state will have a segmented quality. Such extensive development cannot occur, however, until a product group is matured enough to have sufficient sales volume and at least a few stable product designs" [Utterback and Abernathy, 1975, p. 642].

Third, systemic means that processes become highly integrated and developed, which effect even minor changes in other parts of the process. At this stage, when processes are so essential to the system, changes only occur when the market shifts or technological developments are extreme. If businesses largely resist process technology changes, they will experience economic drops or revolutionary changes. In summary, processes go from "fluid" to "automated" and finally "integrated."

Regarding product innovation, Utterback and Abernathy [1975] describe three different patterns in their evolution over time: performance maximizing, sales maximizing, and cost maximizing:

First, we can say that "a firm with a performance-maximizing strategy might be expected to emphasize unique products and product performance, often in anticipation of a new capability that will expand customer requirements" [Utterback and Abernathy, 1975, p. 643].

Second, "sales-maximizing firms would tend to define needs based on their visibility to the customer. Innovations leading to better product performance might be expected to be less likely, unless performance improvement is easy for the customer to evaluate and compare" [Utterback and Abernathy, 1975, p. 644].

Finally, in "the cost-minimizing stage, significant change frequently involves both product and process modifications and must be dealt with as a system. Because investment in process equipment in place is high and product and process change are interdependent, innovations in both product and process may be expected to be principally incremental" [Utterback and Abernathy, 1975, p. 644].

The above-mentioned process and product stages are summarized in Fig. 7.1.

In Fig. 7.1, Utterback and Abernathy [1975] differentiate the patterns that bind the rate of product and process innovations over time. It is noteworthy to observe how the model starts with a maximum rate of product performance and low and uncoordinated process patterns and evolves until the achievement of product cost optimization and process systemization at the end. On the one hand, the authors depict a product innovation cycle characterized by high rates of innovation at the beginning and decreasing over time. This is normal if we think that the major effort in product development is done at the beginning of the product life cycle comprising all the research and development, whereas this effort decreases over time. Accordingly, the model shows how major product innovation is triggered by customer needs, technology, and cost.

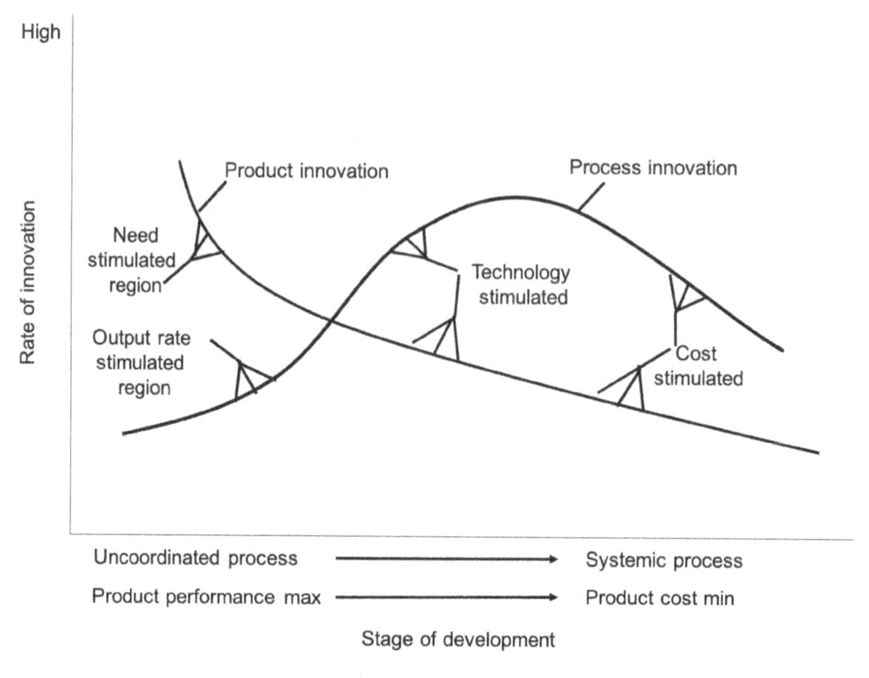

Figure 7.1. Innovation and Stage of Development.
Source: Utterback and Abernathy [1975, p. 645].

On the other hand, process innovations portray a kind of S-curve triggered by the output rate, technology, and costs over time. Whereas, at the beginning, new processes aim to guarantee product development according to product specifications and uniqueness, and in the second stage, technology is used to fulfill demand and maximize sales. Finally, the cost of innovation is triggered at the final stage to optimize processes and products.

Being aware of the necessity of this continuous adaptation and improvement in business processes, the question is as follows: Under what conditions is it expected to initiate the process of change? According to Juran and Godfrey [1999, Section 6.17], the readiness for process change happens when a series of conditions occur: There must be tension for change based on an unsatisfactory perception of the current situation. There must be an attractive or satisfactory alternative based on a vision of how to build a better situation. There must be a step-by-step approach and follow-up system on how to get the desired new state. Once the change is set up, people and the organization itself have to acquire new skills to reach the goals of efficiency and efficacy.

Process development and innovation are crucial for business performance. Accordingly, Davenport [1993], in his work *Process Innovation: Reengineering Work Through Information Technology*, places attention on increasing competition and the changing nature of markets. He referred to the need to go beyond the quality and incremental process improvement approach and embrace new ways of business improvement depending on how a business views itself and how it is structured, pointing out the importance of processes. Therefore, the author claims, the way in which a business is depicted is not in terms of functions, divisions, or products, but in terms of key processes. Consequently, in order to achieve higher levels of improvement, it is necessary to redesign processes from beginning to end, involving technologies and organizational resources, through "process innovation."

7.2.2 *From Operational Improvement to Process Innovation*

To deepen the study of process innovation, it is interesting to delve into its origins. Process innovation is grounded in the necessity to improve efficiency in organizations and it is rooted in different managerial approaches, such as operational improvement approaches, the quality movement, the socio-technical school of thought, various innovation theories, and the use of IT for competitiveness. We are going to examine each of these references to get a detailed understanding of the role of processes in the firm.

Undoubtedly, the first references to systematic structuring of the firm to achieve operational improvement, gain efficiency, and optimize results stem from the seminal principles of *Scientific Management* [Taylor, 1911]. Taylor's was the first attempt to apply scientific principles to organizational processes and business management, and he was considered to have mounted the first serious attempt to develop business process innovation. Taylor's principles mainly address the implementation of systematic scientific techniques to get productivity and efficiency gains. It was the first clear explanation of a systematic and organized management procedure. Thus, scientific management is about the rational systematization and streamlining of human behavior with the aim of discovering the methodical processes hidden behind human practice, the establishment of ideal processes under the patterns of scientific management principles, seeking efficiency and productivity, and guaranteeing standard operating procedures.

However, operational improvement is only the first step and it evolves to more ambitious organizational objectives over time. In this regard, according to the work of Davenport [1993], we present Fig. 7.2, which illustrates the history of innovation evolution, distinguishing between the inspection, continuous improvement, and process innovation phases.

Consequently, product inspection (see, for example, Taylor [1911], or the quality control approach) is the easiest way to seek process improvement, through the simple operationalization of business procedures, though this approach doesn't trigger many changes in the existing processes. Continuous improvement (see, for example, Shewhart and Deming [1986]; Feigenbaum [1951]) involves a higher level of process change. This approach can be proved by quality function deployment (QFD) and total quality management (TQM) procedures. Finally, Davenport [1993] refers to process innovation as the stage where the level of change is higher, which will be addressed later on in this chapter.

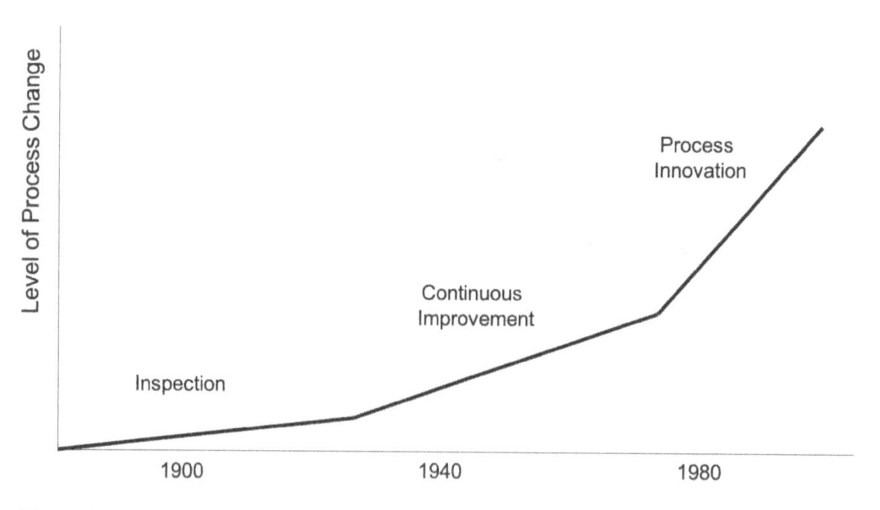

Figure 7.2. Evolution of Innovation.

Source: Davenport [1993, p. 320].

Coming back to the continuous improvement approach, we can say that it stems from the quality movement. The quality approach has its origins in Japan, where it was first introduced to Toyota's Hino Motor in 1975 with impressive results in the automotive industry, being later introduced into the whole Toyota group [Chan and Wu, 2002, p. 465]. As it is raised in his book *Step-by-Step*, Terninko [1997] argues that since the 1960s, quality control and quality improvement lead the manufacturing function in Japan, focusing on process design under the quality approach and promoting what is known as QFD.

The quality management approach has had a clear impact on process improvement, defining it as process quality management (PQM) which is measured by three principal dimensions: effectiveness, efficiency, and adaptability. Juran and Godfrey [1999, Section 6.1] explain the three elements in the following statement: "The process is effective if the output meets customer needs. It is efficient when it is effective at the least cost. The process is adaptable when it remains effective and efficient in the face of the many changes that occur over time." Therefore, quality management opens three ways to process improvement and innovation: the focus on the customer's needs, the optimization and cost-effective orientation, and adaptability, which leads the changing nature of the process, through a continuous adaptation to the customer's needs and qualifying QFD as the prelude to business process innovation.

More particularly, the orientation of PQM is launched with the selection of key processes in the firm, the identification of people and teams responsible of the process, and the establishment of goals and mission statements. The PQM methodology runs through three phases: planning, transfer, and operational management. Process improvement begins in the planning phase, since the process design or redesign is undertaken involving five steps [Juran and Godfrey, 1999, Section 6.6] as follows:

(1) defining the present process;
(2) determining customer needs and process flow;
(3) establishing process measurements;
(4) conducting analyses of measurement and other data;
(5) designing the new process, where the output is the new process plan.

Continuing with the different approaches in the study of business processes in the firm, we refer to the socio-technical school (e.g., Trist and Bamforth [1951]; Lewin [1951]), that points out that process innovation has been inherited as an emphasis on the interplay between technology and people. This approach maintains that technology and humans are enablers of change, and simultaneous management of both aspects is required, although special attention is allowed for the human element in innovation. Creating structured work processes by combining technology and people while taking culture and the environment into account is the goal of this approach.

Now that we have addressed the different approaches in the study of process development, it is worth mentioning that there is a two-fold approach to business process development, which can be addressed from the improvement perspective or from the innovation perspective [Utterback and Abernathy, 1975; Gopalakrishnan *et al.*, 1999; Reichstein and Salter, 2006]. Correspondingly, the first step to process improvement is the process development that includes the identification and evaluation of the process potential, which is carried out within the logical framework of design and control. The final aim of process development is the optimization of the defined framework [Schallmo *et al.*, 2018].

Consequently, many authors have focused on the necessity of managing processes within the firm, which entails process design, control, and development. In this regard, we define business process management as the modeling, execution (including automation), and evaluation of the different processes of the firm [Zur Muehlen, and Indulska, 2010]. Likewise, process management is a means to control and improve the process organization [Biazzo, and Bernardi, 2003]; improve the quality of products and services [Sandhu, and Gunasekaran, 2004]; aligning business processes with strategy of the firm's and customer's needs [Lee and Dale, 1998], and improving effectiveness and business performance [Armistead *et al.*, 1999].

Process management can be seen in a three-fold way: as a structured systematic approach to analyze and continually improve the process; as a holistic view of process management as a part of managing the whole organization; or as a set of tools and techniques for improving processes [Palmberg, 2009, p. 210]. This improvement orientation is supported by

Frishammar *et al.* [2013, p. 215], who define process development as a "deliberate and systemic development related to production objectives, implying the introduction of new elements into the production process with the purpose of creating or improving methods of production," assuming that this process development attempts to improve the production process, is organizationally complex and spans multiple functions, affects other processes or activities, is often internal to the firm, and presupposes the introduction of new elements into the production process. Likewise, Hinterhuber [1995, p. 65, cited in Schallmo *et al.*, 2018] is more focused on the systemic orientation of processes, stating, "by process management we mean a customer oriented management endeavor to achieve exceptional performance in those business processes which transcend functional boundaries. A business process represents a set of integrated and coordinated activities required for producing goods or offering services. The starting point for process management is a precise definition of the business processes. These business processes must be incorporated within the strategy of the firm or within the strategy of particular business."

Notwithstanding the importance of process management to increase business results, the fact is that this approach is more oriented to improvements rather than to innovation. In this vein, it is also interesting to go one step further and refer to the concept of "business process reengineering (BPR)" that is defined by Hammer and Champy [1993, p. 49] as a "fundamental rethinking and radical redesign of business processes to achieve dramatic improvements in critical, contemporary measures of performance, such as cost, quality, service, and speed." Therefore, business reengineering aims not only at incremental changes or adaptations in order to gain efficiency but also at a "rethinking" of or a major shift in the way a process is designed and implemented, where the expected outcome of BPR is "dramatic." Similarly, Ozcelik [2010, p. 2] insists on the same determinant characteristic of BPR and outlines it as "a radical redesign of processes in order to gain significant improvements in cost, quality, and service." Again, to qualify as BPR, the shift ought to be "radical" and the expected improvements "significant." In the same vein, Holland *et al.* [2005] refer to BPR as implementing the methodologies required to change internal business processes and responding to environmental change requirements or internal needs. The authors argue that it is important to know how to analyze current business processes in terms of their ability to meet business objectives and environmental needs. According to Juran and Godfrey [1999, Section 6.3], BPR should be considered part of the family of methodologies within the realm of PQM. The authors state that "Business Process Reengineering (BPR) accomplishes a shift of managerial orientation from function to process. According to the consultants who first described BPR and gave it its name, BPR departs from the other methodologies in its emphasis on radical change of processes rather than on incremental change. Furthermore, BPR frequently seeks to change more than one process at the same time."

Being aware of the aforesaid, different levels of change in business processes can be said to range from merely operational adjustments to improvements, reengineering, and innovations. Accordingly, Frishammar *et al.* [2013] refer to process innovation and process improvement as two ideal types of process development, where process innovation is a major shift or change in business processes. The authors argue that process development is a matter of continuum, where process improvement "refers to continuous improvements of existing processes with the aim of achieving improved efficiency" and process innovation "presupposes all-encompassing changes to the manufacturing process such as the launch of a next-generation process for a new product" [Frishammar *et al.*, 2013 p. 215]. Utterback and Abernathy [1975, p. 641] discussed process innovation focusing on the production process and defining it as a "system of process equipment, work force, task specifications, material inputs, work and information flows, etc. that are employed to produce a product or service." Conversely, with the innovation approach, these authors adopt more of an improvement approach, arguing that process development is linked to productivity improvements over time in an evolutionary pattern, being more intensive in capital, task specialization, rationalization of flows, product standardization, and larger process scales.

7.2.3 *Process Innovation: Definition and Types*

At this point, it is worth focusing on the definition of "process innovation." This concept can, again, be analyzed under a dual perspective: result-oriented or process-oriented. First, result-oriented process innovations are those that are introduced within the company or in the market to improve the firm's economic performance. Under this approach, we have to consider that process innovation varies depending on the innovation object (product or process), level of innovation (incremental or radical), and the reference unit (company-, customer-, or competition-oriented). Second, the process-oriented perspective refers to innovation process "as a sequence of activities and decisions that lead to the marketing of a new product or the use of a new process, connected in a logical and temporal relationship." In this case, the innovation's objectives are the different elements of the process, where the innovation level is also discriminated into radical and incremental, and the primary unit of reference is the customer [Schallmo *et al.*, 2018, p. 24].

Davenport [1993] also refers to process innovation as the performance of business activities in a radically new way to attain dramatic and visible results to meet the business objectives. Again, Davenport qualified process changes as process innovations when both degree of novelty (radically new way) and expected results (dramatic and visible) are outstanding. Also, Reichstein and Salter [2006, p. 1] define process innovation as "new elements introduced into an organization's production or service operations — input materials, task specifications, work and information flow mechanisms, and equipment used to produce a product or render a service — with the aim of achieving lower costs and/or higher product quality." In the same regard, Reichstein and Salter [2006] argue that process innovations are mainly related to the introduction of equipment and machinery and the existence of "learning by doing" and "learning by using" experiences in order to improve productivity. Kline and Rosenberg [2010] stress that system or process innovations are not usually perceived as science-based innovations; but in many cases, they really are the "purest of pure science." Likewise, the authors argue that in the commercial realm, process innovation plays a more important role than science in cost reduction and system performance.

Davenport [1993, p. 10] clearly distinguishes between process improvement and process innovations. Consequently, the author argues that "process innovation can be distinguished from process improvement, which seeks a lower level of change. If process innovation means performing a work activity in a radically new way, process improvement involves

Table 7.1. Process Improvement versus Process Innovation.

	Improvement	**Innovation**
Level of change	Incremental	Radical
Starting point	Existing processes	Clean slate
Frequency of change	One time/continuous	One time
Time required	Short	Long
Participation	Bottom-up	Top-down
Typical scope	Narrow, within functions	Broad, cross-functional
Risk	Moderate	High
Primary enabler	Statistical control	Information technology
Type of change	Cultural	Cultural/structural

Source: Davenport [1993, p. 10].

performing the same business process with slightly increased efficiency or effectiveness."

In Table 7.1, we can see the main differences between process improvement and process innovation [Davenport, 1993, p. 11].

Analyzing the differences between process improvement and process innovations, we can see how the first and sharpest disparities relate to the level of change. While process improvement is incremental, process innovation is categorized as radical. The term *radical* normally refers to the existence of something new, which is not based on an evolution of something already existent and that now has been changed or altered in a specific way. Also, the starting point for process improvement is the existing process, which is analyzed and modified in a way that enhances its efficiency or output, while the triggering point for process innovation is a completely clean slate. Concerning the frequency of change, process improvements are developed under a continuous approach, while process innovation is a one-time shot to reach its desired results.

Clear differences are reported as well concerning the time required for improvement versus innovation. Process improvements are developed on a short-term basis, since they are produced by continuous adaptation and change. However, innovation usually requires more time, since it lasts until the new process is established and running in the business. Some differences exist regarding the participation within the business, with a bottom-up view for the improvement approach, whereas a top-down view is deployed for process innovation. This is reasonable if we consider that people working directly with the firm's processes can report flaws and suggest improvements that can be implemented quickly, while innovation could demand the participation of top-level managers of the firm who have the competence and authority to trigger major changes in processes.

Also, process improvement is characterized by a narrow scope, mainly focused on the task and activities that are scrutinized to be modified into a more effective or productive way; whereas process innovation has a wider focus, where the point is seeking new ways of doing work. Likewise, we can say that the level of risk is low in the improvement approach and higher in the innovation one. Finally, process improvement is mainly enabled by statistical controls that monitor the tool for flaw detection, while information and communication technology (ICT) is the primary enabler of process innovation. All the aforementioned attributes provoke cultural changes in the case of process improvement, where more profound changes regarding process innovation drive both cultural and structural changes.

To complete the characterization of business process innovation, it is also worth referring to the *Oslo Manual* in its 4th edition OECD [2018]. In particular, OECD [2018] states that the term *business process* "includes the core business function of producing goods and services and supporting functions such as distribution and logistics, marketing, sales, and after-sales services; information and communication technology (ICT) services to the firm, administrative and management functions, engineering and related technical services to the firm, and product and business process development" [OECD, 2018, p. 72]. In particular, a business process innovation is defined by *Oslo Manual* as "a new or improved business process for one or more business functions that differs significantly from the firm's previous business processes and that has been brought into use in the firm" [OECD, 2018, p. 72] and "business process innovation" can involve the improvement of a single business function or a combination of different ones and that this can be adopted by the firm itself or by external contractors. By the concept of *improved business function*, the *Oslo*

Manual refers to greater efficacy, resource efficiency, reliability and resilience, affordability, and convenience and usability for those involved in the business process, either external or internal to the firm. The *Oslo Manual* also addresses the motivation of implementing new or improved business processes, such as the implementation of business strategies, cost reduction, improving the quality of products or working conditions, or to meet regulatory requirements.

To clarify the processes and functions that are considered to determine the kind of process innovations within the firm, the 4th edition of the *Oslo Manual*, citing Brown [2008], distinguishes between six process categories. These categories and a detailed description of each of them are listed in Table 7.2.

The *Oslo Manual* (OECD 2018) also notes that many times business process and product innovations are bundled, pointing out the broad character of business innovations. For instance, the manual exemplifies that a business process innovation could also improve the quality of a product, developing both product and business process innovation. Likewise, the implementation of a product innovation could also need a business process innovation, which is the case of a new online function for selling information products that require ICT and web development and at the same time a service innovation for potential users [OECD, 2018, p. 76]. Services are a particular case of business process and product bundling, since production, delivery, and consumption occur simultaneously.

In summary, the OECD [2018, p. 242] defines business process innovation as "a new or improved business process for one or more business functions that differs significantly from the firm's previous business processes and that has been brought into use by the firm. The characteristics of an improved business function include greater efficacy, resource efficiency, reliability and resilience, affordability, and convenience and usability for those involved in the business process, either external or internal to the firm. Business process innovations are implemented when they are brought into use by the firm in its internal or outward-facing operations." Accordingly, *Oslo Manual* stresses the importance of novelty in implementing new business processes in the firm, since it is necessary that these new business processes differ significantly from the previous ones. At the same time, the OECD [2018] points out that business process innovation has a clear goal aimed at improving business performance.

Table 7.2. Functional Categories for Identifying the Type of Business Process Innovations.

Short term	Details and subcategories
1. Production of goods or services	Activities that transform inputs into goods or services, including engineering and related technical testing, analysis, and certification activities to support production.
2. Distribution and logistics	This function includes the following: (a) transportation and service delivery; (b) warehousing; (c) order processing.
3. Marketing and sales	This function includes the following: (a) marketing methods including advertising (product promotion and placement, packaging of products), direct marketing (telemarketing), exhibitions and fairs, market research, and other activities to develop new markets; (b) pricing strategies and methods; (c) sales and after-sales activities, including help-desks, other customer support, and customer relationship management activities.
4. information and communication systems	The maintenance and provision of information and communication systems, including the following: (a) hardware and software; (b) data processing and database; (c) maintenance and repair; (d) web-hosting and other computer-related information activities. These functions can be provided in a separate division or in divisions responsible for other functions.
5. Administration and management	This function includes the following: (a) strategic and general business management (cross-functional decision-making), including organizing work responsibilities. (b) corporate governance (legal, planning, and public relations). (c) accounting, bookkeeping, auditing, payments, and other financial or insurance activities. (d) human resources management (training and education, staff recruitment, workplace organization, provision of temporary personnel, payroll management, and health and medical support). (e) procurement (f) managing external relationships with suppliers, alliances, etc.

(Continued)

Table 7.2. (*Continued*)

Short term	Details and subcategories
6. Product and business process development	Activities to scope, identify, develop, or adapt products or a firm's business processes. This function can be undertaken in a systematic fashion or on an *ad hoc* basis and be conducted within the firm or obtained from external sources. Responsibility for these activities can lie within a separate division or in divisions responsible for other functions, e.g., production of goods or services.

Source: OECD [2018, p. 73] based on Brown [2008].

7.3 THE IMPORTANCE OF ICT FOR PROCESS INNOVATION

7.3.1 *ICT and Business Competitiveness*

ICTs are among the most pervasive and relevant technologies, having a great impact on society and particularly in the businesses realm. ICT has changed the way firms operate through the implementation of new business processes that were unconceivable before its irruption. At the same time, ICT has enabled the design and creation of new products and services, new ways of distribution, and innovative business models that disrupted markets and business sectors. The fact is, nowadays, information and communications technology are modifying production systems, business process redesign, data and information processing, and collaborative teamwork [Uhlenbruck *et al.*, 2003; Akhavan *et al.*, 2006]. Even in the 1990s, Morton [1996, p. 158] emphasized the importance of ICT and stated, "information technology is a critical enabler of the re-creation (redefinition) of the organization. This is true in part because it permits the distribution of power, function, and control to wherever they are most effective, given the mission and objectives of the organization." Therefore, process innovation and business transformation based on ICT have been frequent topics of discussion in the technological and business realm. In this section, we will analyze the role of information technologies in business processes and the role of these technologies as enablers of business innovation.

The implementation of information technologies throughout history corresponds to multiple phases or stages. As we addressed earlier in Chapter 1 of this book, the age of communication and information technologies corresponds with the fifth technological revolution [Perez, 2010], which came after four previous major technological transformations (industrial, steam and railway, steel, electricity and heavy engineering, and oil, automobile, and mass production). Perez [2010] refers to the fifth technological revolution as the "Age of Information and Telecommunications." This era is characterized by what is called the "information revolution, which includes microelectronics, computers, software, telecommunications, control instruments, computer-aided biotechnology, and new materials technologies in the field." There is a common agreement to qualify ICT as a general-purpose technology (GPT) [Jovanovic and Rousseau, 2005] because of its pervasiveness (the GPT should spread to most sectors), its improvement (increase of performance and price lowering at the same time), and its capacity for innovation spawning (easing the invention and production of new products and processes) [Bresnahan and Trajtenberg, 1995].

According to Porter and Heppelmann [2014], information technologies have radically shaped the competitive environment on two occasions. They contend that we are on the verge of a third major change based on IT applications. The authors observe three IT waves that have impacted the business realm over time. The first one (1960s and 1970s) was focused on the standardization of processes across companies, such as the automation of activities in the value chain, order processing, computer-aided design (CAD), and manufacturing resource planning, with a huge impact on productivity gains. This stage has been characterized as doing things better, looking for more productivity, efficiency, and gains, where IT has played an enabling role.

The second wave (1980s and 1990s) was grounded in the rise of the Internet because it "enabled coordination and integration across individual activities; with outside suppliers, channels, and customers; and across geographies. It allowed firms, for example, to closely integrate globally distributed supply chains" [Porter and Heppelmann, 2014, p. 4]. The first two waves mainly affected productivity and growth, transforming the value chain but separate from major changes in products and services.

Finally, concerning the third wave, the authors say that IT became "an integral part of the product itself. Embedded sensors, processors, software, and connectivity in products (in effect, computers are being put inside products), coupled with a product cloud in which product data is stored and analyzed and some applications are run, are driving dramatic improvements in product functionality and performance. Massive amounts of new product-usage data enable many of those improvements" [Porter and Heppelmann, 2014, p. 4].

The above-mentioned orientation explains the existence of two different types of innovations, either based on internally or externally oriented IT capabilities [Neirotti and Raguseo, 2017]. The difference is that internal capabilities lead to internal information processing by supporting cross-functional integration while external capabilities allow organizations to respond to market variations and changes in the needs of customers and suppliers. On the one hand, internal IT capabilities are mainly aimed at improving production planning and control processes (e.g., scheduling, inventory management) and administrative processes (e.g., billing, accounting, and payroll) with a clear impact on cost savings, efficiency improvements, and operational control. On the other hand, external IT capabilities are aimed at processes such as sales, logistics, and NPD that effect growth, product differentiation, or focalization strategies (one-to-one marketing) and market adaptation. Neirotti and Pesce [2019, p. 383] postulate that "innovations falling in the first type placed cost reduction, efficiency, operational results and business process improvement among the main goals for ICT investments"; while "the second type of innovations support firms in doing new things rather than doing the same things with less." This approach is consistent with the two-fold orientation in process development, differentiating between process improvement and process innovation as we have discussed previously in this chapter.

In particular, some authors highlight that IT is especially relevant to business process innovation, since they are "a collection of technologies capable of translating business process models into computer-supported activities, relinquishing routine management and control tasks from the organizational agents" [Antunes and Mourão, 2011, p. 1241]. In the same vein, Pyon *et al.* [2011, p. 3268] emphasize the supportive character of

ICT, referring to them as a "system which supports business processes using methods, techniques, and software to design, enact, control and analyze operational processes involving humans, organizations, applications, documents and other sources of information." The *Oslo Manual* [2018] also highlights the importance of information technologies. Specifically, the *Oslo Manual* argues that digitalization allows the use of IT in a wide variety of tasks enabling its performance and has the potential to "transform business processes, the economy and society in general" [OECD, 2018, p. 38]. The *Oslo Manual* also recognizes the importance of digitalization for both product and process innovation and states, "Digital technologies and practices are pervasive across business processes. They are used to codify processes and procedures, add functions to existing processes and enable the sale of processes as services. The implementation of business process innovations is therefore often tied to the adoption and modification of digital technologies" [OECD, 2018, p. 72].

7.3.2 *Impact of ICT on Process Innovation*

Concerning the influence of ICT, Bassellier and Benbasat [2004] defend that information systems (IS) and information technology (IT) have heavily impacted business innovation, particularly for Internet-based organizations, since IT enables process innovation while businesses require a deep understanding of IT competences and resources [Sambamurthy *et al.*, 2003]. There is also sectoral evidence of how the use of IT enhances innovation, in the electronics industry (microprocessors) [Thomke, 2006] and retail industry [Pantano, 2014].

Process innovation has also been traditionally enabled by information technologies, which can support process reengineering or change, and it has been widely addressed in the research literature. This is the case for software for process design and simulation [Serrano and Den Hengst, 2005], the existence of flexible infrastructures [Broadbent *et al.*, 1999], and the use of project management (PM) systems [Attaran, 2003]. Likewise, Kogut and Zander [1992] argue the importance of IS to enhance the combinative ability to improve knowledge management processes. Also, Al-Mashari and Zairi [2000] conducted research that shows a relationship between IT infrastructure and BPR. Accordingly, authors found that businesses view IT infrastructure and organizational infrastructure as mutually evolving, where the information needs of new processes determine both IT infrastructure and IT capabilities. In this vein, the socio-technical perspective seems to be more suitable for implementing successful BPR. For instance, the use of document management, databases, and communication networks are the most widely implemented technologies used to enable BPR today [Grover *et al.*, 1995]. In the same regard, Hengst and Vreede [2004, p. 89] report that the use of modeling and collaboration tools facilitates BPR. The authors argue that "simulation models describe processes in the organizational system with graphical symbols, as conceptual models do, and provide quantitative information about these processes. Simulation modeling tools provide a structural environment in which one can understand, analyze and improve business processes."

Again, regarding the impact of IT in the business realm, we can say that there is evidence in the literature that demonstrates the positive correlation of IT implementation and business performance. Accordingly, Ollo-Lopez and Aramendia-Muneta [2012] argue that the adoption of information and communication technologies has a positive effect on productivity, boosts innovation, and enhances productivity and business performance. Similarly, literature reports that the use of ICT improves efficiency, effectiveness, and competitiveness [Johnston *et al.*, 2007; Mahmood *et al.*, 2000]; productivity growth [Black and Lynch, 2001; Matteucci *et al.*, 2005]; innovation performance of the business [Gërguri-Rashiti *et al.*, 2017]; competitiveness gains [Anand *et al.*, 2013]; enhanced competitive advantage, improved customer satisfaction, and the ability to enable business innovations [Lyytinen and Rose, 2003; Peppard and Ward, 2004]. Also, it has been reported that ICT reduces the number of workers — production workers in particular — as a result of automation in many business practices, such as supply chain management, order management, and customer service management. At the same time, ICT eases the process of subcontracting to reduce costs and improve efficiency [Dube *et al.*, 2007].

It is also interesting to cite the work of Tarafdar and Gordon [2007], which highlights the importance of IT competencies in business process innovation. Thus, the authors present the following competences as key enablers of business process innovations:

- *Knowledge management*: The management of knowledge is critical to the success of the innovation process (see Chapter 6 for further references). Innovation is a knowledge-intensive process and the existence of knowledge management systems favors the process itself. In this vein, the authors argue that "a competency in knowledge management requires IT resources such as expert systems and data mining software for knowledge creation and abstraction, database systems for storage and retrieval, portals for knowledge dissemination, and decision support systems for knowledge application" [Tarafdar and Gordon, 2007, p. 362].
- *Collaboration*: The authors maintain that a competency in collaboration triggers innovation at the initial stages of the innovation process and it helps the work to develop an innovative idea. Thus, IT supporting collaboration can ease the innovation process itself. The authors maintain that "information technology, such as communication networks, email, webcams, file-synchronization software, multi-user editors, wikis, encryption software, and portals can help an organization develop a competency in collaboration. Cross-company collaboration, in particular, demands such a competency" [Tarafdar and Gordon, 2007, p. 363].
- *Project management*: PM competence is critical to innovate, since firms can improve cross-functional integration and planning [Thieme *et al.*, 2003]. Some authors point out the importance of technology in PM. For instance, Tarafdar and Gordon [2007, p. 363] argue, "As an IS competency, PM draws on hardware and software for storing and manipulating product and process data, and software for scheduling and tracking tasks, assigning and monitoring staffing, and documenting and evaluating progress at milestones."
- *Ambidexterity*: Ambidexterity is a concept that points out the ability of the organization to balance both strategic vision and operational excellence, which also includes the capacity to balance the response to internal and external needs. The concept of ambidexterity has been extended to the IS function, describing "an ambidextrous systems

development unit as one that has organizational structures, processes and tools required" [Tarafdar and Gordon, 2007, p. 364]. The authors refer to this ambidexterity as the necessity that the IS function covers the supply aspects (building and maintaining basic infrastructure such as servers, databases, and networks and running them efficiently and reliably at a low cost) and the demand aspects (communicating with business unit leaders, identifying business problems and opportunities, experimenting with different solutions, risk taking, and creating business accountability for IT projects).

- *IT/Innovation governance*: The authors define IT/innovation governance as the management mechanisms including authority and communication among innovation and IT managers. IT/innovation governance involves IT infrastructure planning and security, central IT activities and facilities, and collaboration between IT and innovation activities. Accordingly, suitable governance mechanisms facilitate innovation.
- *Business–IT linkage*: Business strategy and IT need to be closely linked in order to integrate ITs into a product or a process innovation. IT professionals have to work closely with other business areas in order to comprehend operations and needs, so that this function can provide solutions and innovations in different fields and areas of business.
- *Process modeling*: "A process modeling IT competency is a relatively low-level competency that expresses facility and experience in the use of these tools and techniques for modeling processes. A process modeling IT competency supports process innovation by providing structure to the redesign process and cognitive support for those involved in it" [Tarafdar and Gordon, 2007, p. 366].

It seems to be assured that IT impacts business process innovation, but we have to be aware that its influence can reach different stages depending on the level of use and integration of IT in the business. In this regard, Venkatraman [1994] linked the use of ICT in the firm dominion with the transformation of the business, arguing that there are different levels of IT-enabled business transformation, from evolutionary levels (localized exploitation — internal integration) to revolutionary levels (business process redesign — business network redesign — business scope redefinition). The localized exploitation level is related to the leveraging ability of IT to redesign business operations. The internal integration level is about leveraging the IT capability to design a seamless organizational process with technical interoperability and organizational interdependence. Finally, business process redesign is about reshaping "key processes to derive organizational capabilities to competing in the future as opposed to simply rectifying current weaknesses; use IT capabilities as enabler for future organizational capability" [Venkatraman, 1994, p. 82]. Similarly, Lee [2012] differentiates two levels of the application of ICT to the business environment: incremental and radical or disruptive changes. The author states that, whereas operational innovations lead to incremental changes in the business, radical transformations are understood as the implementation or creation of a process that is innovative in an industry and that leads to wide changes. In the same regard, Lyytinen and Rose [2003] referred to the importance of disruptive IS innovation based on Internet computing, which has transformed the application portfolio and development practices over time.

It is worth noting how the different stages of business change enabled by IT correspond to the process innovation and process improvement levels of process development referenced earlier [Frishammar *et al.*, 2013]. Accordingly, Dedrick *et al.* [2003] defend that IT is not only a tool for automating existing processes but more importantly an enabler of organizational changes, which includes process innovation. IT enabling process-oriented businesses is also reported by Kohli and Sherer [2002], arguing that a process approach in organizations can positively impact the payoff of IT investments. In the same vein, Peppard and Ward [2004, p. 169] defend that technology is more than an isolated resource but a means to achieve organizational objectives, stating that "technology itself has no inherent value and that IT alone is unlikely to be a source of sustainable competitive advantage. The business value derived from IT investments only emerges through business changes and innovations, whether they are product/service innovation, new business models, or process change, and organizations must be able to assimilate this change if value is to be ultimately realized."

Likewise, considering the importance of IT as an enabler of organizational change, it is worth citing the work of Anaya *et al.* [2015, p. 786], which studied the role of enterprise IS in businesses, reaching the following conclusions. First, they suggest that organizations "should lead their IT projects as business change projects, and not installation of technological-based projects; because many technology projects like ERP, CRM, or any other EIS projects come up with many opportunities, which should be exploited, such as introducing innovative processes or reengineering existing processes." Second, that the approach of the implementation of ICT as a business change project requires the adoption "of change and innovation in their mind-set and in their culture."

Anaya *et al.* [2015, p. 775] stress the importance of IT in leveraging business innovation. The authors argue that information technologies and systems enable organizations "not only [to] automate their business processes but also redesign, improve and introduce innovative processes, taking the benefit from the systems that work across functional departments and from the communications infrastructure embedded in the systems. The role of the enterprise systems here is not only to automate the business processes, but also to introduce deep changes about how the work is done." In the same vein, Peppard and Ward [2004] applied the resource-based theory approach to information system management and proposed an approximation called innovation-based approach that highlights how ICT resources and capabilities could be used to improve business and raise organizational innovation. This approach is summarized by Anaya *et al.* [2015] in the following terms:

Mode 1: Enabling the implementation of something new that couldn't be done before the appearance of the IT application. In this regard, IT is imperative to drive business innovation, which will not be possible without it. The authors consider that this kind of innovation can be categorized as radical.

Mode 2: Using IT to do something in a new way, like "the introduction of a new process or practice to do the existing work but in new ways introduced and developed by existing IT resources" [Anaya *et al.*, 2015, p. 774], considering these kinds of innovations as incremental.

Mode 3: Doing something new with the use of IT applications, such as the development of new products, services, or even processes, where IT applications play a role in support of the innovation process itself.

Also, Lee *et al.* [2011] defend that ICT changes business processes in three ways. First, through the use of ICT in business, hierarchies flatten and at the same time the degree of centralization is lowered. The use of ICT eases the sharing and exchange of information and access to managers and enhances a manager's information processing capacity and control [Pan and Jang, 2008; Karmarkar and Vandana, 2004]. Second, ICT impacts workplace BPR, since it enables changes in the coordination process, it brings people closer in spite of distances, and it enables more work and coordination between geographically dispersed business units [Morton, 1996; Dube *et al.*, 2007; Sarkar and Singh, 2006]. Finally, ICT provokes workforce changes, reducing the number of errors as a result of automation in different business practices.

To end with an analysis of the importance of information and communication technologies and its impact on business process innovation, it is also interesting to make reference to the trending technologies that are shaping business. In this regard, we cite the work of Martin and Leurent [2017], who in the report "Technology and Innovation for the Future of Production: Accelerating Value Creation," depict the major technologies that are shaping what is known as the Fourth Industrial Revolution. The authors argue that "advances in the internet of things, artificial intelligence, advanced robotics, wearables and 3D printing are transforming what, where and how products are designed, manufactured, assembled, distributed, consumed, serviced after purchase, discarded and even reused" [Martin and Leurent, 2017, p. 4]. The report presents a production technology radar made of seven major technological areas: human–machine interface, connectivity and computing, analytics and intelligence, digital physical transformation, production philosophies, advance materials, and advance production processes. Each of the aforementioned technologies is classified as mainstream technologies and mature and emerging technologies. Table 7.3 displays the aforementioned classification.

Table 7.3. Production Technology Radar.

Technological area	Emerging technologies	Maturing technologies	Mainstream technologies
Human–machine interface		Wearable devices (with AR/VR capability); co-robotics; conversational systems.	Social networks; context-based systems; multimodal interaction; intuitive UIs; dialogue systems; AR/VR.
Connectivity and computing	Quantum computing, "Smart dust"	Blockchain; quantum communication; quantum cryptography; adaptive security architecture; interoperability.	Apps and platforms; mobile Internet; cloud computing; modeling, simulation, and visualization; M2M connectivity; digital twin.
Analytics and intelligence	Cognitive computing	Bioinformatics; knowledge-based automation; embedded cognitive functions; deep learning.	Remote maintenance; intelligent systems; numerical modeling and algorithms; data mining; knowledge-based systems; big data.
Digital physical transformation	4D printing	Autonomous robotics; collaborative robotics; flexible and reconfigurable machinery and robots.	3D printing; photonics; mechatronics; new machine architectures.
Production philosophies	Product life cycle management for advanced materials; dynamic manufacturing execution environments.	Flexible, modular manufacturing systems; energy/material/resource efficient manufacturing; integrated product development; green, sustainable production.	Recycled materials; new business models; mass customization; product services.
Advanced materials	Nanoengineering of materials and surfaces; materials for 4D printing.	Printed electronics; flexible electronics; perovskite solar cells; meta-materials; 3D molding; multi-scale/multi-material manufacturing.	High-value ceramics; lightweight materials; semiconductors; coatings, surfaces, and layers; composite materials; biotechnology.

(Continued)

Table 7.3. *(Continued)*

Technological area	Emerging technologies	Maturing technologies	Mainstream technologies
Advanced production processes	Integration of non-conventional technologies, nano-assembly	Physical, chemical, and physicochemical processes; manufacturing of high-performance flexible structures; advanced forming, joining, and machining.	Surface manufacturing processes; continuous manufacturing; net and near net shape manufacture; inkjet printing; manufacturing of biofuels.

Source: Martin and Leurent [2017, p. 7].

REFERENCES

Akhavan, P., Jafari, M., and Fathian, M. [2006]. Critical success factors of knowledge management systems: A multi-case analysis, *European Business Review*, 18(2), pp. 97–113.

Al-Mashari, M., and Zairi, M. [2000]. Revisiting BPR: A holistic review of practice and development, *Business Process Management Journal*, 6(1), pp. 10–42.

Anand, A., Wamba, S. F., and Gnanzou, D. [2013]. A literature review on business process management, business process reengineering, and business process innovation, *Workshop on Enterprise and Organizational Modeling and Simulation*, Springer, pp. 1–23.

Anaya, L., Dulaimi, M., and Abdallah, S. [2015]. An investigation into the role of enterprise information systems in enabling business innovation, *Business Process Management Journal*, 21(4), pp. 771–790.

Antunes, P., and Mourão, H. [2011]. Resilient business process management: Framework and services, *Expert Systems with Applications*, 38(2), pp. 1241–1254.

Armistead, C., Pritchard, J. P., and Machin, S. [1999]. Strategic business process management for organizational effectiveness, *Long Range Planning*, 32(1), pp. 96–106.

Attaran, M. [2003]. Information technology and business-process redesign, *Business Process Management Journal*, 9(4), pp. 440–458.

Bassellier, G., and Benbasat, I. [2004]. Business competence of information technology professionals: Conceptual development and influence on IT-business partnerships, *MIS Quarterly*, pp. 673–694.

Biazzo, S., and Bernardi, G. [2003]. Process management practices and quality systems standards: Risks and opportunities of the new ISO 9001 certification, *Business Process Management Journal*, 9(2), pp. 149–169.

Black, S. E., and Lynch, L. M. [2001]. How to compete: The impact of workplace practices and information technology on productivity, *Review of Economics and Statistics*, 83(3), pp. 434–445.

Bresnahan, T. F., and Trajtenberg, M. [1995]. General purpose technologies "Engines of growth"? *Journal of Econometrics*, 65(1), pp. 83–108.

Broadbent, M., Weill, P., and St. Clair, D. [1999]. The implications of information technology infrastructure for business process redesign, *MIS Quarterly*, pp. 159–182.

Brown, S. P. [2008]. Business processes and business functions: A new way of looking at employment, *Monthly Labor Re*view, 131(12), pp. 51–70.

Chan, L. K., and Wu, M. L. [2002]. Quality function deployment: A literature review, *European Journal of Operational Research*, 143(3), pp. 463–497.

Davenport, T. H. [1993]. *Process Innovation: Reengineering Work Through Information Technology* (Harvard Business Press, USA).

Dedrick, J., Gurbaxani, V., and Kraemer, K. L. [2003]. Information technology and economic performance: A critical review of the empirical evidence, *ACM Computing Surveys (CSUR)*, 35(1), pp. 1–28.

Dube, P., Liu, Z., Wynter, L., and Xia, C. [2007]. Competitive equilibrium in e-commerce: Pricing and outsourcing, *Computers & Operations Research*, 34(12), pp. 3541–3559.

Dumas, M., La Rosa, M., Mendling, J., and Reijers, H. A. [2013]. *Fundamentals of Business Process Management* (Springer, Heidelberg).

Feigenbaum, A. V. [1951]. *Quality Control: Principles, Practice and Administration: An Industrial Management Tool for Improving Product Quality and Design and for Reducing Operating Costs and Losses* (McGraw-Hill, USA).

Frishammar, J., Lichtenthaler, U., and Richtnér, A. [2013]. Managing process development: Key issues and dimensions in the front end, *R&D Management*, 43(3), pp. 213–226.

Gopalakrishnan, S., Bierly, P., and Kessler, E.H. [1999]. A reexamination of product and process innovations using a knowledge-based view, *Journal of High Technology Management Research*, 1(10), pp. 147–166.

Gërguri-Rashiti, S., Ramadani, V., Abazi-Alili, H., Dana, L. P., and Ratten, V. [2017]. ICT, innovation and firm performance: The transition economies context, *Thunderbird International Business Review*, 59(1), pp. 93–102.

Gold, B. [1987]. Approaches to accelerating product and process development, *Journal of Product Innovation Management*, 4(2), pp. 81–88.

Grover, V., Jeong, S. R., Kettinger, W. J., and Teng, J. T. [1995]. The implementation of business process reengineering, *Journal of Management Information Systems*, 12(1), pp. 109–144.

Hammer, M., and Champy, J. [1993]. Reengineering the corporation: A manifesto for business revolution, *Business Horizons*, 36(5), pp. 90–91.

Harrison, M. I. [2004]. *Diagnosing Organizations: Methods, Models, and Processes*, 3rd edn., Vol. 8, (Sage Publications, USA).

Hengst, M. D., and Vreede, G. J. D. [2004]. Collaborative business engineering: A decade of lessons from the field, *Journal of Management Information Systems*, 20(4), pp. 85–114.

Hinterhuber, H. [1995]. Business process management: The European approach, *Business Change and Reengineering*, 2 [4], pp. 63–73.

Holland, C. P., Shaw, D. R., and Kawalek, P. [2005]. BP's multi-enterprise asset management system, *Information and Software Technology*, 47(15), pp. 999–1007.

Johnston, D. A., Wade, M., and McClean, R. [2007]. Does e-business matter to SMEs? A comparison of the financial impacts of internet business solutions on European and North American SMEs, *Journal of Small Business Management*, 45(3), pp. 354–361.

Jovanovic, B., and Rousseau, P. L. [2005]. General purpose technologies. In *Handbook of Economic Growth*, Aghion, P. and Durlauf, S. (eds.), (Elsevier, Amsterdam, The Netherlands), Vol. 1, pp. 1181–1224.

Juran, J., and Godfrey, A. B. [1999]. *Quality Handbook*, 5th edn. (McGraw-Hill, USA).

Karmarkar, U., and Vandana, M. [2004]. *Business and Information Technologies Annual Report* (UCLA Anderson School of Management, USA).

Kline, S. J., and Rosenberg, N. [2010]. *Studies on Science and The Innovation Process*, Rosenberg, N. (ed.), Chapter 11, *An Overview of Innovation* (World Scientific, USA) pp. 173–203.

Kogut, B., and Zander, U. [1992]. Knowledge of the firm, combinative capabilities, and the replication of technology, *Organization Science*, 3(3), pp. 383–397.

Kohli, R., and Sherer, S. A. [2002]. Measuring payoff of information technology investments: Research issues and guidelines, *Communications of the Association for Information Systems*, 9(1), pp. 241–268.

Lee, O. K. D. [2012]. IT-enabled organizational transformations to achieve business agility, *The Review of Business Information Systems* (Online), 16(2), p. 43.

Lee, R. G., and Dale, B. G. [1998]. Business process management: A review and evaluation, *Business Process Management Journal*, 4(3), pp. 214–225.

Lee, Y. C., Chu, P. Y., and Tseng, H. L. [2011]. Corporate performance of ICT-enabled business process re-engineering, *Industrial Management & Data Systems*, 111(5), pp. 735–754.

Lewin, K. [1951]. *Field Theory in Social Science: Selected Theoretical Papers*, Cartwright, D. (ed.), (Harper & Row, New York, USA).

Lim, L. P., Garnsey, E., and Gregory, M. [2006]. Product and process innovation in biopharmaceuticals: A new perspective on development, *R&D Management*, 36(1), pp. 27–36.

Llewellyn, N., and Armistead, C. [2000]. Business process management: Exploring social capital within processes, *International Journal of Service Industry Management*, 11(3), pp. 225–243.

Lyytinen, K., and Rose, G. M. [2003]. Disruptive information system innovation: The case of internet computing, *Information Systems Journal*, 13(4), pp. 301–330.

Mahmood, M. A., Mann, G. J., and Zwass, V. [2000]. Impacts of information technology investment on organizational performance, *Journal of Management Information Systems*, 16(4), pp. 3–10.

Martin, C., and Leurent, H. [2017]. *Technology and Innovation for the Future of Production: Accelerating Value Creation*, World Economic Forum, Geneva Switzerland.

Matteucci, N., O'Mahony, M., Robinson, C., and Zwick, T. [2005]. Productivity, workplace performance and ICT: Industry and firm-level evidence for Europe and the US, *Scottish Journal of Political Economy*, 52(3), pp. 359–386.

Mooney, J. G., Gurbaxani, V., and Kraemer, K. L. [1996]. A process oriented framework for assessing the business value of information technology, *SIGMIS Database*, 27(2), pp. 68–81.

Morton, M. S. [1996]. How information technologies can transform organizations, *Computerization and Controversy: Value Conflicts and Social Choices*, (Morgan Kaufmann, San Diego, USA), pp. 148–160.

Morris, D. C., and Brandon, J. [1993]. *Re-engineering Your Business* (McGraw-Hill, USA).

Neirotti, P., and Pesce, D. [2019]. ICT-based innovation and its competitive outcome: The role of information intensity, *European Journal of Innovation Management*, 22(2), pp. 383–404.

Neirotti, P., and Raguseo, E. [2017]. On the contingent value of IT-based capabilities for the competitive advantage of SMEs: Mechanisms and empirical evidence, *Information & Management*, 54(2), pp. 139–153.

OECD and Eurostat [2018]. Oslo Manual 2018. Guidelines for Collecting, Reporting and Using Data on Innovation, 4th Edition; The Measurement of Scientific, Technological and Innovation Activities (OECD publishing Paris/ Eurostat, Luxembourg).

Ollo-López, A., and Aramendía-Muneta, M. E. [2012]. ICT impact on competitiveness, innovation and environment, *Telematics and Informatics*, 29(2), pp. 204–210.

Ozcelik, Y. [2010]. Do business process reengineering projects payoff? Evidence from the United States, *International Journal of Project Management*, 28(1), pp. 7–13.

Palmberg, K. [2009]. Exploring process management: Are there any widespread models and definitions? *The TQM Journal*, 21(2), pp. 203–215.

Pan, M. J., and Jang, W. Y. [2008]. Determinants of the adoption of enterprise resource planning within the technology-organization-environment framework: Taiwan's communications industry, *Journal of Computer Information Systems*, 48(3), pp. 94–102.

Pantano, E. (2014). Innovation drivers in retail industry. *International Journal of Information Management*, 34(3), pp. 344–350.

Peppard, J., and Ward, J. [2004]. Beyond strategic information systems: Towards an IS capability, *The Journal of Strategic Information Systems*, 13(2), pp. 167–194.

Perez, C. [2010]. Technological revolutions and techno-economic paradigms, *Cambridge Journal of Economics*, 34(1), pp. 185–202.

Pisano, G. P. [1997]. *The Development Factory: Unlocking the Potential of Process Innovation* (Harvard Business Press, USA).

Porter, M. E., and Heppelmann, J. E. [2014]. How smart, connected products are transforming competition, *Harvard Business Review*, 92(11), pp. 64–88.

Pyon, C. U., Woo, J. Y., and Park, S. C. [2011]. Service improvement by business process management using customer complaints in financial service industry, *Expert Systems with Applications*, 38(4), pp. 3267–3279.

Reichstein, T., and Salter, A. [2006]. Investigating the sources of process innovation among UK manufacturing firms, *Industrial and Corporate Change*, 15(4), pp. 653–682.

Sambamurthy, V., Bharadwaj, A., and Grover, V. [2003]. Shaping agility through digital options: Reconceptualizing the role of information technology in contemporary firms, *MIS Quarterly*, pp. 237–263.

Sandhu, M. A., and Gunasekaran, A. [2004]. Business process development in project-based industry: A case study, *Business Process Management Journal*, 10(6), pp. 673–690.

Sarkar, A. N., and Singh, J. [2006]. E-enabled BPR applications in industries banking and cooperative sector, *Journal of Management Research*, 6(1), pp. 18–34.

Schallmo, D. R., Brecht, L., and Ramosaj, B. [2018]. *Process Innovation: Enabling Change by Technology: Basic Principles and Methodology: A Management Manual and Textbook with Exercises and Review Questions,* 1st edn. (Springer Gabler, Germany).

Serrano, A., and Den Hengst, M. [2005]. Modeling the integration of BP and IT using business process simulation, *Journal of Enterprise Information Management*, 18(6), pp. 740–759.

Shewhart, W. A., and Deming, W. E. [1986]. *Statistical Method from the Viewpoint of Quality Control* (Courier Corporation, USA).

Talwar, R. [1993]. Business re-engineering — A strategy-driven approach, *Long Range Planning*, 26(6), pp. 22–40.

Tarafdar, M., and Gordon, S. R. [2007]. Understanding the influence of information systems competencies on process innovation: A resource-based view, *The Journal of Strategic Information Systems*, 16(4), pp. 353–392.

Taylor, F. W. [1911]. *The Principles of Scientific Management New York* (Harper & Brothers, New York, USA).

Teece, D. J., Pisano, G., and Shuen, A. [1997]. Dynamic capabilities and strategic management, *Strategic Management Journal*, 18(7), pp. 509–533.

Terninko, J. [1997] *Step-by-Step QFD: Customer-Driven Product Design,* 2nd edn. (CRC Press, US).

Thieme, R., Michael Song, X., and Shin, G. C. [2003]. Project management characteristics and new product survival, *Journal of Product Innovation Management*, 20(2), pp. 104–119.

Thomke, S. H. [2006]. Capturing the real value of innovation tools, *MIT Sloan Management Review*, 47(2), pp. 24–32.

Trist, E. L., and Bamforth, K. W. [1951]. Some social and psychological consequences of the Longwall Method of coal-getting: An examination of the psychological situation and defenses of a work group in relation to the social structure and technological content of the work system, *Human Relations*, 4(1), pp. 3–38.

Uhlenbruck, K., Meyer, K. E., and Hitt, M. A. [2003]. Organizational transformation in transition economies: Resource-based and organizational learning perspectives, *Journal of Management Studies*, 40(2), pp. 257–282.

Utterback, J. M., and Abernathy, W. J. [1975]. A dynamic model of process and product innovation, *Omega*, 3(6), pp. 639–656.

Utterback, J. [1994]. *Mastering the Dynamics of Innovation: How Companies Can Seize Opportunities in the Face of Technological Change* (Harvard Business School Press, USA).

Venkatraman, N. [1994]. IT-enabled business transformation: From automation to business scope redefinition, *Sloan Management Review*, 35, pp. 73–73.

Yoshida, S. [1998]. Development and innovation in the Japanese chemical industry, *International Journal of Technology Management*, 15(6–7), pp. 568–585.

Zairi, M. [1997]. Business process management: A boundaryless approach to modern competitiveness, *Business Process Management Journal*, 3(1), pp. 64–80.

Zur Muehlen, M., and Indulska, M. [2010]. Modeling languages for business processes and business rules: A representational analysis, I*nformation Systems*, 35(4), pp. 379–390.

CHAPTER 8

THE DEVELOPMENT OF NEW PRODUCTS

8.1 INTRODUCTION

The development and launching of new products and services to the market is essential for firm survival and the growth and prosperity of the firm, in particular to the firm's profitability and competitiveness, since successful new products can provide a sustainable competitive advantage and lead the firm itself to overall success [Loch *et al.*, 1996]. According to Edgett's [2011] benchmark study from the American Productivity and Quality Center (APQC), new products launched in the last 3 years account for 27.3% of company sales, which indicates that the product portfolio is renewing constantly. Nowadays, in a fast-paced, evolving market, adaptation to new customers' needs is a critical factor for the success of companies.

Furthermore, there is a huge variety of literature that has investigated the relationship between product innovation and firm development [Klette and Griliches, 2000; Klette and Kortum, 2004], where the development of new products has a relevant impact on company performance [Booz *et al.*, 1982].

Companies are facing major challenges with a dramatic reduction in product life cycles and an increase in the heterogeneity of the customers in the market, which pushes both manufacturing and service firms to improve the development of new products to compete in the market. Marketplace in recent times is depicted by high rates of competitive pressures, rapid changes in customer preferences, the high pace of technological evolution, and shorter product life cycles [Menon *et al.*, 2002]. Manufacturers are trying to introduce new products to the market rapidly, even though speed is not always the best option [McNally *et al.*, 2011]. Thus, new product development (NPD) is a key element of the innovation strategy of the firm that maintains an updated portfolio to fulfill a compelling market.

There is a wide range of literature on NPD as this topic has drawn the interest of scholars and practitioners for decades [Mahajan and Wind, 1988, p. 1993; Cooper and Kleinschmidt, 2007]. The study of NPD has been approached from different points of view as well, such as engineering [Perrone *et al.*, 2010], sustainability [Gmelin and Seuring, 2014; Baumann *et al.*, 2002], globalization [Townsend *et al.*, 2010], and collaboration and networks [Johnsen, 2009; Coviello and Joseph, 2012].

However, firms have to face an inconvenient reality, which is that the majority of new products never even get into the market, and those that do suffer a high failure rate [Crawford, 1987]. There is seminal literature that shows results in this vein. For instance, broad industry cross-section research conducted by the Product Development Management Association (PDMA) in a total of 189 firms in 1990 showed that 11 ideas were needed to obtain one successful product, reporting a success rate in new product launches of 58% [Page, 1993]. The well-known research of Booz *et al.* [1982] reported that 4 out of 7 ideas enter the development process, but only 1.5 are launched and only 1 becomes a success. The aforementioned study of Edgett [2011] follows in the same vein, reporting that only half of NPD projects achieve their financial objectives and only 44.4% of them are launched on time in the market. Therefore, the central question in the study of NPD is as follows: Why do some products fail while others succeed and what are the key elements in determining a winning innovative product?

8.2 WHAT IS NEW PRODUCT DEVELOPMENT?

First, we have to be aware that NPD itself is a knowledge creation and sharing process [Iansiti and MacCormack, 1997]. Following the definition posed by the *Oslo Manual*, we can say that a product innovation is "a new or improved good or service that differs significantly from the firm's previous goods or services and that has been introduced on the market" [OECD and Eurostat, 2018, p. 21]. The manual specifies that "product innovation must provide significant improvements to one or more characteristics or performance specifications. This includes the addition of new functions, or improvements to existing functions or user utility. Relevant functional characteristics include quality, technical specifications, reliability, durability, economic efficiency during use, affordability, convenience, usability, and user friendliness" [OECD and Eurostat, 2018, p. 71].

PDMA posits that the term new product has a multifaceted approach, but it can be generally defined as a "product (either a good or service) new to the firm marketing it. Excludes products that are only changed in promotion." NPD can be referred to as "the overall process of strategy, organization, concept generation, product and marketing plan creation and evaluation, and commercialization of a new product." In the same vein, NPD Process is defined as a "disciplined and defined set of tasks and steps that describe the normal means by which a company repetitively converts embryonic ideas into salable products or services" [Kahn, 2012].

Therefore, NPD consists of all the activities carried out by the firm to develop and launch new products to the market. NPD encompasses a number of multifaceted and interrelated activities that start from the generation and assessment of new product ideas and the incorporation of consumer requirements to the specification of new products and launching and marketing activities among others [Hilletofth and Eriksson, 2011]. Usually, NPD starts with the identification of a market opportunity and reference to a set of technologies that can jointly be integrated into a viable new product [Marion *et al.*, 2012].

According to Booz *et al.* [1982], the development of new products entails different stages starting with an idea or product concept, which is then evaluated, developed, tested, and launched into the market. However, the NPD process is not homogeneous across firms or industries, but should be adapted to the specific requirements of the business or sector.

8.3 NEW PRODUCT DEVELOPMENT MODELS

There has been wide interest in studying the structure and phases of the development of new products. NPD is a key process and its optimization could result in a significant increase in new product success and, therefore, in great impact on business performance. Nevertheless, NPD models are aimed at minimizing risk in the process of putting new products into the market. Therefore, many researchers have proposed different models to explain and subsequently manage NPD.

In the literature, we have found models that are based on breaking down the NPD process into different stages or activities, models that seek overlapping and simultaneous ways to improve the effectiveness of the process, and finally, models interacting with the external environment and networks. We studied earlier, in Chapter 2, the innovation process including all the activities that lead the company to generate innovative outputs and we also have done a first approach to modeling innovation. Once we assume that innovation can be studied as a process, we can approach the management of innovation as the conscious process of organization, control, and execution of all the activities that compose it. We are specifically interested in delving into the development of new products to manage it properly.

According to Cooper [1990, p. 3], the new product process is a "formal blueprint, roadmap, template, or thought process for driving a new product project from the idea stage through to marker launch and beyond."

Wagner [2012] argues that according to NPD literature, the NPD process is usually divided into two phases, the fuzzy front end (FFE), where the firm conducts early pre-development work (idea generation to project evaluation), and a second phase in a more formal and structured NPD process, which is carried out with project management methods. The first phase is a less structured, non-routine oriented, *ad hoc* decision process that is highly dynamic and has a high level of uncertainty [Zhang and Doll, 2001].

Following Zhang and Doll [2001, p. 177], the NPD process normally includes a number of components among which we can cite the following:

- *Use of a Structured Development Process (SDP)*: This is a set of different activities that comprise the NPD process, setting the rules of the game and specifying how to enter the process, through major milestones, schedules, and resource assignment criteria.
- *The Review Board*: Normally composed of senior managers and executives who resolve issues apparent in the overall process.
- *Realization Teams*: Normally a cross-functional execution team which champions the success of a product. These teams are often called Integrated Program Teams (IPTs) or Core Teams and they report to the Review Board.

- *Phase (or Stage) Gate Reviews*: These are considered the spine of the project; they signify the main milestones in the NPD process.

All models available in the literature emphasize the same feature, that is, the structured character of the NPD process [Kleinschmidt and Cooper, 1991; Cooper and Kleinschmidt, 1995; Filippini *et al.*, 2004; Troy *et al.*, 2008; Cooper and Edgett, 2008] and the benefits of utilizing these kinds of models in the development of new products [Cooper and Kleinschmidt, 2007]. The complex nature of the innovation phenomena makes it difficult to establish a common and general NPD innovation process.

8.3.1 *The Activity-Stage Model*

Perhaps the most commonly known model that is used to describe the NPD process is the activity-stage model [Trott, 2008; Kotler *et al.*, 2002], where the basic steps are essentially the same and only the stage or phase names differ from one model to the other. One of the seminal and best known is from Booz *et al.* [1982]. These authors studied the role of management in creating and exploring new products, including in their analysis more than 700 American manufacturers and their NPD practices, which gathered information on 13,000 products — both industrial and consumer goods. They distinguished six phases in the NPD processes: new product strategy development, idea generation, screening and evaluation, business analysis, development, testing, and commercialization. These authors discovered that those businesses that follow a systematic approach to the development of new products are more likely to send successful goods and services to the market. In what follows, we briefly describe each of the stages:

- *New product strategy development*: In this stage, the company seeks to determine the products of interest to the company according to the strategy of the firm. This can be done by scrutinizing major company problems, checking the availability of resources, pinpointing growth opportunities to expand the growth of the firm, and identifying technological breakthroughs that may increase profit margins.
- *Idea generation*: In this stage, the company establishes an idea generation program, identifying the sources of ideas and involving

groups of people. Once the firm has identified the sources of generation, it has to give them a clear concept of the company's interests and inspire creative people to generate ideas. Accordingly, a team approach is recommended, as well as minimizing distractions from short-term problems in the firm. In this stage, it is also necessary to design ways of collecting ideas through an organized network (i.e., establishing collection points), launching idea collection procedures, organizing the outside input of ideas, actively asking for new ideas, and considering ideas on a can-do basis.

- *Screening*: The interest of the company in this stage is going from the idea to the product concept; that is, driving ideas into key business implications. To do so, it is useful to gather data and opinions from customers about the business proposition. Idea selection techniques are used in this stage to be efficient with the number of ideas to work on. One of the key points is being able to appraise each idea for the potential it has in the market, revenues and profit estimations, needed investments, and risk assessment.
- *Business analysis*: This stage is focused on proposing the idea to the people responsible for evaluating it (i.e., a team representing major departments involved in its development). Furthermore, the company has to delve into the determination of the desirable market features for the product and its feasibility, market trends, competitor's comparisons, market and technical research, and identifying product characteristics that could help to accelerate the product's introduction into the market. At the end of this stage, the firm has to be able to develop specifications and set up a final plan for the product (final business plan, timetables, expenditures, and top management approval).
- *Development*: Each product has to be developed by the firm through a comprehensive plan by breaking down the product proposal into different projects for control management (scheduling, team staffing, responsibilities, and major milestones). It also requires that products are built according to available information (security, market studies, and specifications). Finally, it is necessary to conduct laboratory evaluation and prepare the product for testing, which comprises lab tests, product appraisal, commercialization tests, and final management report on the product.

- *Testing*: This stage includes commercial experiments and tests to check for product acceptance in the market, conducting testing by users in real-life situations and production and market tests. After that, the company has to make final product decisions (test findings, program for full-scale production, and a final product and commercialization program submission to top management).
- *Commercialization*: This is the final stage, which includes complete final plans for production and marketing and the initiation of coordinated production and selling programs. After the initial implementation, it is crucial to evaluate results and make continuous improvements and adjustments to sales and production.

Kotler *et al.* [2002] go further with the discussion of the activity-based model of NPD, proposing a nine-stage path, quite similar to the Booz *et al.* [1982] model. The model breaks down the process into the stages of product strategy, idea generation, idea screening, concept development and testing, marketing strategy, business analysis, product development, marketing test, and commercialization.

8.3.2 *Stage-Gate Idea Model*

Nevertheless, one of the most cited and recognized models to explain the NPD process is the stage-gate idea to launch process, which was originally proposed by Cooper [1990 and 1994]. This author claims that many leading firms (DuPont, 3M, HP, Procter and Gamble (P&G), Northern Telecom, ICI-UK, IBM, Dow Chemical, Polaroid, Black & Decker, and Exxon Chemicals) had developed a systematic stage-gate process, which is a roadmap from idea to launch consisting of discrete stages, where each stage is preceded by a Go/Kill decision point or gate [Cooper, 1990, p. 4]. Cooper says that the Stage-Gate systems recognize product innovation as a process. By doing so, it is assumed that innovation, and more particularly, product innovation can be managed, applying process management methodologies to the innovation process itself.

The author posed what he coined as the third-generation process with overlapping, fluid stages and "fuzzy" or conditional Go decisions at Gates, which include substantial modifications of the traditional stage-gate process. According to the author, the third-generation new product process model is more cross-functional (no one stage is owned by any specific function); has a more holistic orientation, capturing all the processes from idea to launch and not only the central processes of NPD; has a much stronger market orientation of the model itself, the customer being an integral part of the product development process; introduces a parallel concurrent processing where it is not necessary to completely conclude the previous phase to step forward to the next stage; and finally, it incorporates sharper decision points with clear Go/ Kill — not just control mechanisms as in earlier models. Summing up, the author claims third-generation models to be fluid (with overlapping and fluid stages for greater speed), fuzzy (featuring conditional decisions dependent on the situation), focused (looking at the overall portfolio and prioritizing the best ideas), and flexible (considering each project as unique).

The stage-gate system divides the process into different stages, each of them containing a number of activities that sometimes overlap. Cooper [1990] proposed that a typical system could involve four to seven stages depending on the company or division that is running it. Another characteristic is that each stage is more costly to execute than the preceding one, and at the same time, the risk goes down with each stage since the information becomes better with every step. Each gate is characterized by a series of inputs or deliverables, a set of exit criteria, and a defined output. A typical stage-gate system is depicted in Fig. 8.1.

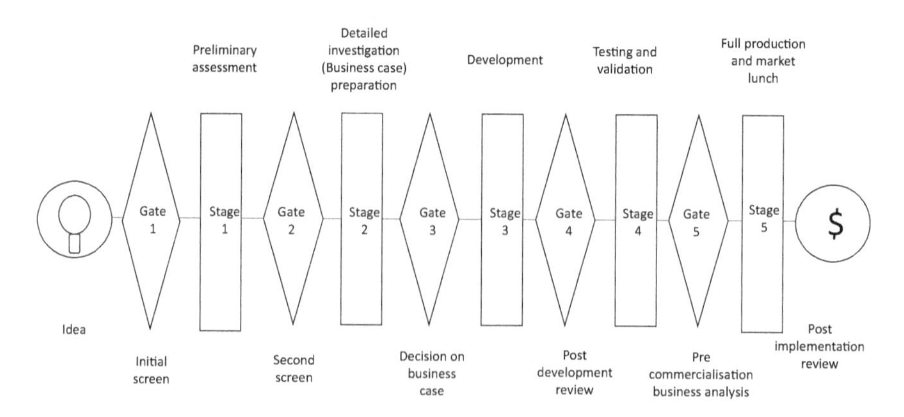

Figure 8.1. An Overview of a Stage-Gate System.

Source: Cooper [1990].

This figure, though generic, describes a typical stage-gate system skeleton. Let's define the different stages according to Cooper [1990] as follows:

A typical stage-gate process starts with the idea for a new product that is submitted to Gate 1, the initial screen. This is the starting point of the project and the first decision made here is whether to allocate resources to a project or not. If the decision is positive (Go), the project moves to the "Preliminary Assessment Stage." The criteria used in this stage deals with strategic alignment, project feasibility, magnitude of the opportunity, differential advantage, synergy with the firm's core business and resources, and market attractiveness.

- *Stage 1: Preliminary Assessment* — This stage aims at evaluating the technical and market strengths of the project. Preliminary market evaluations involve a series of activities such as secondary sources research, getting in contact with key users, focus groups, and screening concept tests with potential users. The objective is to determine potential market size and likelihood of acceptance.
- At the same time, a preliminary technical assessment is being developed through an in-house appraisal of the product (development and manufacturing feasibility and costing). This screening evaluation of both market and technical issues can be deeply evaluated at Gate 2.
- *Gate 2*: At this stage, a new re-evaluation of the project is made, essentially with the same criteria used in Gate 1, though taking into account the new information gathered in the previous stage. Besides the "must" and "should" criteria used in the previous stage, sales force and customer reaction to the proposed product are also considered at this stage. The result of this stage is a checklist that facilitates the decision. Financial issues are considered, but only in a simple way (i.e., payback period).
- *Stage 2: Definition* — This is the stage where the project is defined in a more detailed way, prior to further and heavy spending and moving to product development. At this stage, the consumer's needs have to be determined through market research and concept testing to define the characteristics of a winning product. Afterwards, necessities from the market must be translated into technically and economically viable solutions, developing some lab work but not as far as a full development

project. As a result of this stage, an operations appraisal is conducted taking into account issues such as manufacturing, investment, or legal and patent aspects if required. A full financial analysis is also developed as input to Gate 3.

- *Gate 3: Decision on Business Case* — At this stage, the project is on the verge of entering the development stage, where it takes heavy spending. It is the last occasion to turn the project down before significant resources are committed. The project again considers "must" and "should" criteria and contains a revision of the Stage 2 evaluation and a detailed evaluation of key items before the project moves to Gate 3. This evaluation entails market definition, product concept, positioning strategy, products benefits, desired product features, attributes, and specifications. Development, preliminary operations, and marketing plans are checked and approved at this gate.
- *Stage 3: Development* — At this stage, the product is being developed and at the same time detailed marketing and operations plans are created. Financially, the project is deeply analyzed, and legal, copyright, and patent issues are agreed.
- *Gate 4: Post-development Review* — With the same logic of all the models, the previous development phase is reviewed again. At this gate, a deep financial analysis is executed with more accurate data. Once this gate is approved, plans are ready for implementation in the next stage, while marketing and operations plans are reviewed for future implementation.
- *Stage 4: Validation* — This is a key stage where the whole viability of the project is tested, encompassing the product itself, production processes, customer acceptance, and the financial and economic aspects of the whole project. In a more detailed way, the following tasks are undertaken in this stage: In-house product tests through which quality assurance and performance are being evaluated; user field trials of the product, to check that the functions work well in use conditions; pilot production to determine costs and rates in a more specific way; market pre-test and tests, as well as selling trials to foresee the possible market share and revenues, and, ultimately, the economic viability of the project.
- *Gate 5: Pre-commercialization Decision* — This is the final stage before commercialization of the product and the last chance to stop the

whole project before going to market. Following stage-gate logic, the quality of the validation stage is reviewed. Financial issues are key to giving a "yes" to move to the next stage, while marketing and operations are approved for implementation in the next stage.

- *Stage 5: Commercialization* — In this stage, market and operations plans are implemented.
- Finally, there is a post-implementation review. When the new product becomes a regular product in the firm's portfolio and its performance is reviewed. Thus, we evaluate costs, expenditures, profits, and timing in comparison to forecasts. A critical assessment of the strengths and weaknesses of the project is carried out.

8.4 SUCCESS FACTORS IN NPD

Studying NPD as a process is an interesting approach, since it allows us to measure the effectiveness of different practices in the development of new products. Consequently, identifying these good practices and adopting them in our business could help us replicate the NPD success of the best performing companies [Dooley *et al.*, 2002]. In much the same way, both the PDMA's [Barczak *et al.*, 2009] and the American Productivity Quality Center [Cooper *et al.*, 2004a, 2004b, 2004c] have conducted best practices analyses in the field of NPD. It is also worth citing the work of Kahn *et al.* [2006] who explored the underlying dimensions of the NPD domain, developing a framework for analysis for NPD best practices. These authors suggest that there are six dimensions of NPD domains for the study of best practices: strategy, portfolio management, process, market research, people, and metrics and performance. In what follows, a description of the best practices related to each of these fields is provided.

Firms that approach NPD from a strategic point of view choose a long-term vision and are more focused on identifying future market needs. According to Cooper *et al.* [2002], strategically oriented firms are more likely to identify customer needs and new product strategies and they consider this a good way of defining their NPD approach. It seems that the effective management of NPD starts early in the FFE phase and not only in the more structured and managed development process. Accordingly, this FFE determines the success of NPD projects. Thus, Langerak *et al.* [2004] provided evidence that market orientation is related positively to product advantage and to proficiency in market testing, launch budgeting, strategy, and tactics. Conversely, market orientation has a direct relationship neither with new product performance nor with organizational performance.

"Portfolio management is a dynamic decision process, whereby a business list of active new products (and R&D) projects is constantly up-dated and revised. In this process, new projects are evaluated, selected and prioritized; existing projects may be accelerated, killed, or de-prioritized; and resources are allocated and re-allocated to the active projects" [Cooper *et al.*, 2006, p. 4]. Concerning portfolio management, Cooper *et al.* [2002] argue that only 21.2% of firms have well-performed product portfolio management and many of them declare that their portfolio management is very weak. Better product portfolio management results in a better allocation of resources and a mix of breakthrough and incremental NPD projects.

Process is related to the different stages of NPD. According to the authors, almost 47% of firms have a clear definition of assessment and evaluation of the project at each stage, and 40% name a process manager in order to ensure the correct implementation of it. The results show that best performing companies are those who have implemented a clearly defined process with stages and gates.

Also, market research is shown as a best practice that is incorporated by organizations to maintain the customer's point of view throughout the phases of the NPD process. Market research encompasses all activities to learn from customers, including changes in the environment. These techniques include concept, product, and market testing in relation to the gates and stages defined in the NPD process. Leading companies in that field allocate resources to identify unarticulated customer needs, customer response to proposed products, price sensitivity, and market potential, among others.

The people dimension includes human resources and team issues related to the NPD process. The authors point out that about 42% of companies use champions to lead the NPD efforts. Likewise, cross-functional teams are important in the NPD process, jointly with central NPD functions that are involved full-time in managing this task.

Finally, we refer to the metrics of the process itself. All these activities relate to how the process is measured, tracked, reported, recognized, and rewarded. Best performing companies have clear, properly defined Go/Kill criteria for each step of the NPD process. Furthermore, these firms are more likely to identify customer satisfaction, time to market, market value, etc.

Kahn *et al.* [2012] conducted empirical research including a Delphi methodology and a survey in the US, UK, and Ireland to gather opinions about the importance of the different dimensions for the development of new products, based on their previous work [Kahn *et al.* 2006]. As a result of this investigation, the authors pinpoint seven dimensions (strategy, process, culture, project climate, research, metrics, and commercialization) of interest in comparison to six from the aforementioned framework. Strategy was considered the top, followed by research, commercialization, and process, while project climate and metrics were noted as less important.

The authors identified best and worst practices in each of the seven dimensions. Table 8.1 shows the best practices for each of them.

Cooper [2011] identifies nine critical success factors in NPD. Correspondingly, winning companies in the NPD arena first have a product innovation and technology strategy, which enhances ideation, roadmap, and resource allocation. Second, these companies set a clear focus on developing fewer, better, and the right mix of projects, adopting systematic portfolio management on a Go/Kill basis. Third, companies leverage core competences that are needed for NPD, instead of stepping out with projects unrelated to core competences of the firm, when collaborative projects and open innovation projects lower the risk of the firm. Fourth, those projects aimed at attractive markets perform better and should be among the selection criteria. Fifth, allocation of resources is a key part in the success of NPD. A lack or scarcity of resources is a clear predictor of project failure. Sixth, the right organizational structure supports NPD and influences the success of new products. Seventh, a corporate culture supporting and fostering innovation also makes business excel in NPD. Eighth, top management support adds to the success of NPD. Finally, the implementation of a stage-and-gate launch system positively impacts the development of new products.

Table 8.1. Best Practices of NPD for Each Dimension.

Dimensions	Best practices
Strategy	Clearly defined and organizationally visible NPD goals/long-term vision on NPD/aligning NPD with strategic plan/NPD projects' continuous revision/opportunity identification linked to strategic plan.
Process	A common NPD process/Go/Kill criteria clearly defined/NPD process flexibility/clear NPD process and procedures/the NPD process is robust.
Culture	Top management supports the NPD process/management rewards.
Project climate	Cross-functional teams underlie the NPD process/formal and informal coordination of NPD activities.
Research	Market research to identify market needs/concept, product, and market testing/customer/user is an integral part of the NPD process/formally evaluated testing results.
Commercialization	Cross-functional launch team/after-project retrospective meetings/logistics and marketing closely related/customer service and support are part of the launch team/launch process exists.

Source: Kahn *et al.* [2012].

Cooper [2013] delved into the analysis of the aforementioned success factors highlighting the following:

- *Product Innovation and Technology Strategy*: The author, based on Cooper and Edgett [2010], states that having an NPD strategy is linked to positive performance and pinpoints the following ingredients to ensure success: clearly defined product innovation goals and objectives, the role of product innovation in achieving overall business goals, strategic arenas defining areas of strategic focus on which to concentrate new product efforts, employing strategic buckets, having a product roadmap in place, and having long-term commitment.
- *Focused and Sharp Project Selection Decisions — Portfolio Management*: It is argued that many companies suffer from a dispersion of efforts in NPD that stems from a lack of project evaluation and prioritization. Thus, based on the work of Cooper and Edgett [2006], the author posed some important success drivers in this domain such

as strategic alignment, competitive and product advantage, market attractiveness, leverage, technical feasibility, and risk and return.

- *Leveraging Core Competencies — Synergy and Familiarity*: The point is that NPD must be approached from a position of strength, trying to achieve synergies with the core competences and resources of the firm. Taking the idea a little bit further, the author states that the key elements to take into consideration must be, "R&D or technology resources (for example, ideally the new product should leverage the business' existing technology competencies); marketing, selling (sales force), and distribution (channel) resources; brand, image, and marketing communications and promotional assets; manufacturing or operations capabilities and resources; technical support and customer service resources; market research and market intelligence resources; [and] management capabilities" [Cooper, 2013, p. 22].
- *Targeting Attractive Markets*: The author considers two factors: on the one hand, market potential and, on the other hand, the competitive situation in the market. Taking into account the first element, growing market share in a clear market segment and defined customer needs are desired; while focusing on the competitive situation, the author defines negative markets as those that are price-related and have competitors with strong sales force and channels and customer support.
- *The Necessary Resources*: When a project suffers from a lack of interest, support, or resources, the expected result is higher failure rates. When companies try to do more with less and they cut resources, they are jeopardizing the success of the development of new products. Based on Cooper and Edgett [2003], the author affirms that the best developing firms commit necessary resources to NPD and outdo other firms. Equally important, points out the author, is the involvement of team members in many projects and cross-functional product innovation groups.
- *The Way Project Teams Are Organized*: According to the author and based on Nakata and Im [2010] and Cooper [2011], how the project is organized influences the project's results. The author reports that best-performing businesses organize teams, clearly assigning a team to each NPD project, identifying the people who belong to the team and do any work for it. At the same time, people who participate in the

project team do not merely represent the function they primarily belong to, but they belong to the team and assume specific tasks in it. Second, the project team remains stable until the project ends, not just for a fixed period of time or only for a single phase. Third, there is a clear project leader who is in charge of the project from beginning to end. Fourth, the project has a central, shared information source for all team members. Finally, the project team's results are accountable, differentiating the best from the worst performers.

- *The Right Environment — Climate and Culture*: The proper climate and culture for NPD success include risk-taking behavior based on entrepreneurial spirit in the business, acceptance of the top management of risk, investing in occasional risky projects, assessment and recognition of team performance, avoiding micromanagement and second-guessing team members, and open character of review meetings.
- *Top Management Support*: This is a key point for NPD success, setting the stage for product innovation and making a long-term commitment with product innovation as a means of growth, making available necessary resources and establishing NPD processes. Top management must be also committed to the evaluation and selection phase.
- *A Disciplined, Multi-stage, Idea-to-Launch System*: Many researchers have reported a positive effect of the use of a systematic Idea-to-Launch system in the NPD process, improving efficiency and their rate of success [Edgett, 2011; Cooper, 2011; Lynn *et al.*, 1999].

We have studied the most important factors influencing NPD success. Nonetheless, as we have mentioned before, there are some key elements that are particularly relevant to the NPD process, driving the business to better success rates and higher efficiency. Yet, as NPD is a multifaceted process, it is necessary to delve into some aspects to shed light on it.

For instance, it is interesting to refer to the participation of customers in product development processes. Morgan *et al.* [2018] studied customer participation in NPD in 243 firms of different sizes in 14 countries. They found that customer participation is positively related to NPD performance and that innovativeness appears to play a mediating effect in this relationship. Authors also report that these effects are contingent on the absorptive capacity of the firm, since businesses with high absorptive capacity are able to engage customers more than those with lower levels, especially in the latter stages of the NPD process. These authors made an intensive literature review of previous research on customer participation in NPD. Thus, they cite the work of Fang [2008], who argued that there is a trade-off between innovativeness and speed to market, where customer participation improves time to market especially when there is downstream customer network connectivity. Morgan *et al.* [2018, pp. 500–501] conducted a deep literature review in which they referred to the following findings in relation to customer participation: lead customers are a crucial source of ideas for new products [von Hippel, 1976, 1986]; firms develop more new products when customers are closely embedded into the firm [Yli-Renko and Janakiraman, 2008]; customer participation in the ideation and launch stages enhances NPD performance; however, customer participation in the development phase delays time to market and NPD performance [Chang and Taylor, 2016]; customer involvement in NPD improves technical quality and pace of innovation, and consequently, it has an effect on sales and competition [Carbonell *et al.*, 2009]; the sharing of information and coordination effectiveness is positively impacted by customer participation and results in product value improvement [Fang *et al.*, 2008]; customer participation in NPD throughout the co-creation of new products clearly improves the creation of new knowledge that achieves better NPD performance [Mahr *et al.*, 2014]; interaction with customers increases customer information quality in the NPD process and subsequently NPD performance [Bonner, 2010]; the creation of cross-functional NPD teams, failure and reward systems, conflict resolution systems, and product championing have a positive impact on NPD performance [Joshi and Sharma, 2004]; the involvement of customers in NPD results in customer empowerment, better customer orientation, better attitudes toward the firm, and intentions to purchase the new product [Fuchs and Schreier, 2011]; customer participation in NPD improves service quality and firm performance, such as sales, market share, and profits [Ngo and O'cass, 2013]; successful innovations appear to be linked to customer participation in the stages of opportunity recognition, customer-based funding, development and testing, commercialization, and feedback [Coviello and Joseph, 2012]; and radical innovations are positively impacted by the involvement of lead users, mainly referred to as customers leading with new technologies and high motivation [Enkel *et al.*, 2009; Chatterji and Fabrizio, 2014].

Even though the participation of customers seems to be key to NPD, the involvement of other partners also leads the process of innovation to be more networked and open [Utterback *et al.*, 2006]. The necessity to reduce time to market and the growing technological knowledge available induce firms to encourage collaboration with suppliers during the NPD process [Koufteros *et al.*, 2010; Wagner, 2010; Hong *et al.*, 2011].

This openness in the innovation process is known as open innovation and consists of tapping into external ideas brought into the business' innovation process. This means that, apart from the participation of customers, which has been widely studied, other actors such as suppliers also play an important role in the development of innovation, particularly in the NPD process. There are some references that pinpoint the importance of suppliers in the innovation process, where 69% of knowledge sources are reported to come from them [Enkel *et al.*, 2009], and they are the most important and successful source of ideas [Un *et al.*, 2010]. This supplier integration or involvement is materialized in collaboration with the supplier along the NPD process, normally characterized by long-term, partnership-based relationship, with high levels of trust, commitment, and open communication [Monczka *et al.*, 2000; van Echtelt *et al.*, 2007].

Partnering with suppliers is quite often done to conduct the innovation process in all kinds of firms. Frequently, businesses rely on suppliers to access new technologies or knowledge that support the innovation process itself. This can be positive for the firm, since it accelerates the pace of innovation and the speed to enter the market. However, collaboration with suppliers also has inconveniences because this knowledge is available to competitors who can also hire the same sources of knowledge, making it difficult to get any differentiation and consequently building a sustainable competitive advantage [Kessler *et al.*, 2000]. Accordingly, researchers also point out this dual vision of the integration of suppliers into the innovation process.

On the one hand, there is a huge piece of research reporting significant improvements in innovation performance [Hagedoorn, 2002]. At the same time, supplier involvement on teams generally results in a higher achievement of NPD goals and at the same time there is an increase in the sharing of information and knowledge [Petersen *et al.*, 2003]. Moreover, the integration of suppliers improves quality and raises the rate of success of radical NPD in new ventures [Song and Di Benedetto, 2008]. In the case of the FFE and NPD project performance, there is a strong positive relationship within supplier integration [Wagner, 2012]. Furthermore, in environments of technology uncertainty, some elements of the supplier integration process are more likely to be employed, leading to significant improvements in cost, quality, and cycle time objectives. Ragatz *et al.* [2002] and Koufteros *et al.* [2005] collected data from 244 manufacturing firms across several industries to study the internal and external integration of product development, finding that both internal and external integration positively influence product innovation and quality, and ultimately, profitability.

Further, there is also a good array of references that find no or negative effect of supplier integration on NPD performance. Accordingly, Kessler *et al.* [2000] investigated how different technology sourcing strategies throughout NPD influenced innovation speed, development costs, and competitive advantage. They conducted research on 75 NPD processes from large US companies in different industries, finding that more external sourcing during the early stages of NPD was related to lower competitive success, while more external sourcing during later NPD stages lowers innovation speed. The aforementioned study of Koufteros *et al.* [2005] reported that the integration of suppliers in the product development process had adverse effects on quality, especially for a high-complexity environment.

Von Corswant and Tunälv [2002] conducted a case study on determining the critical factors of success in supplier collaboration. The case study was conducted on a Swedish auto manufacturer and five suppliers of the firm. The results showed that the suppliers' internal organization of product development and production and their cooperation with other manufacturers and suppliers were found to be of crucial importance.

Another point refers to the fact that NPD is not an isolated phenomenon in the innovation realm since it is related to other forms of innovation in the business, in particular with innovation in processes. Literature has investigated the closed relationship between product and process innovation; for instance, innovation in processes is usually accompanied by changes in products and vice versa [Lager, 2002; Tang, 2006], discovering that such a relationship will enhance cost efficiency in production, favor product launching, and create new opportunities for product and process innovation [Pisano, 1997; Pisano and Wheelwright, 1995].

Hullova *et al.* [2016] studied the differences in resources and capabilities combined with the specific needs of a new product and process development. The authors developed a classification system defining seven unique complementarities between product and process innovation, illustrating the complementarity map, all taking into account three contingency factors such as technology trajectories, power of supply chain, market potential, and realized absorptive capacity.

There are different approaches to the study of the relationship between product and process innovation. To study the product and process innovation relationship in detail, we refer to the work of Hullova *et al.* [2016], who state that:

- Product innovation creates a need for process innovation. Following the "industry life cycle model," Abernathy and Utterback [1978] argue that changing the rates of product and process innovation depends on the developmental stage of the industry (radical product innovation in the fluid phase, radical process innovation in the transitional phase, and both incremental product and process innovation in the specific patterns phase).
- Product and process innovation are interdependent. This synergetic point of view has been investigated throughout the literature, concluding that it leads to improvements in cost efficiency of production, enhancement in new product launching, easier commercialization, high return on capital, and identifying new opportunities in NPD [Pisano and Wheelwright, 1995; Capon *et al.*, 1992; Milgrom and Roberts, 1995]. The simultaneous execution of both product and process innovation is referred to as the third phase of the industry life cycle theory, in which innovation aims to improve product quality, reduce cost, and gain efficiency by means of process innovation.
- Product and process innovations are two separate kinds of innovation. The authors referred to situations where both process and product innovations run separately. For instance, the acquisition of new

machinery opens up new possibilities for future product innovation, but it doesn't occur simultaneously. There are also situations where lack of capacity of the firm to implement this new technology could hinder process or product innovation.

8.5 THE DEVELOPMENT OF NEW SERVICES

Scholars and practitioners agreed on the growing importance of innovation in services and the study of new services development, since the services sector's contribution to economic growth is widely recognized in developed economies [De Jong and Vermeulen, 2003; Hauknes and Miles, 1996; Miles, 2005]. The difference between products and services has been studied across the literature, making a distinction between the specific characteristics that define both of them.

It is precisely Miles [2005] who argues that services have four specific characteristics, which are intangibility, heterogeneity, inseparability, and perishability, that make services different from goods. Furthermore, the author posits another unique feature of services, pointing out that services offer benefits through access or temporary possession, instead of ownership, with payments taking the form of rentals or access fees. Also, Vargo and Lusch [2004, p. 326] worked on the distinction of goods and services and claim that "goods and services are not mutually exclusive (e.g., tangible versus intangible) subsets of a common domain, that is, products," arguing that the intangibility, heterogeneity, inseparability, and perishability characterizations fail to adequately delineate services from goods.

Thus, there is a growing debate about whether or not service innovation can be treated in the same way as innovation in goods in the manufacturing industry. Coombs and Miles [2000] presented three different orientations to study and define innovation in services. First, there is an assimilation approach where services are treated as goods and similar to the manufacturing approach to innovation. Second, what the authors call a demarcation approach reasoning that service innovation is different from innovation in manufacturing and it has distinctive characteristics. Finally, there is a synthesis approach where some useful elements from manufacturing innovation are being neglected when talking about service innovation.

Going one step further, Drejer [2004] pointed out the peculiarities of services that have led to the development of some new innovation concepts specifically aimed at services. Thus, the first concept in relation to service innovation is *ad hoc* innovation, which was originally introduced by Gallouj and Weinstein [1997], presenting the concept as an interactive social construction to a particular problem posed by a given client. *Ad hoc* innovation becomes an interactive process through which new knowledge and competences are developed to meet the specific needs of the innovation process. Therefore, new development processes in services may differ substantially from physical goods, being aware that in the case of services, innovation intrinsically occurs in the provision of the service.

Another concept in relation to innovation in services is "external relationship innovation," where a network of relationships is established with customers, suppliers, competitors, and other stakeholders to develop innovation in services [Djellal and Gallouj, 2001]. This orientation had been largely studied since the Schumpeterian definition of innovation, where it was referred as the organization of industry and has been gathered in the last version of the OECD Manual [OECD and Eurostat, 2018] as process innovation. Den Hertog *et al.* [2010] referred to the client interface interaction as a way of doing service innovation, which has been widely studied in the literature [Eiglier and Langeard, 1977; Grönroos, 2007; Hauknes and Miles, 1996].

Third, Drejer [2004] refers to the concept of "formalization innovation," which is described as the process aimed at putting service characteristics into order, specifying, and making it concrete [Gallouj and Weinstein, 1997]. Sometimes, the formalization process can be related to a service solution [Leiponen, 2002], which can be defined as a predefined service product as opposed to the service provider functioning as an outside expert. Following the same line, the codification of knowledge is a key point for the formalization of the knowledge applied to the development of a service.

Finally, the author refers to the concept of "expertise-field" innovation, which Gallouj [2000] defined as the process of detecting new needs and responding to them through the process of accumulating knowledge and expertise. The difference between "expertise innovation" and *ad hoc* innovation is that the first one is focused on the openness of new markets, diversification, renewal of product ranges, and creation of a competitive advantage.

There are authors like Howells *et al.* [2004], who posit the differences between innovation patterns in goods and services. Studying the innovation process in services, they state that over 70% of manufacturers focus their innovation efforts on developing new products and only 8% were concentrated on organizational changes. By contrast in services, only 35% identify product or process innovations as their main innovation activity, while 37% were focused on organizational change. Therefore, the innovative effort seems to be higher in goods than in services, but at the same time, non-technological innovations seem to be more likely in services than in physical products. Also, the OECD and Eurostat [2005, p. 11], in its third edition, recognizes that innovation in services may differ substantially from innovation in industrial goods or the manufacturing-oriented sector, affirming that innovation in services "is often less formally organized, more incremental in nature, and less technological." It seems this applies not just to the differences between goods and services but also among the different components of the service sector. Accordingly, Tether and Metcalfe [2004] refer to the diverse character of the service sector, defining a framework of services that includes four types: services engaged in physical transformation of goods (i.e., transport, handling, and storage of goods), services engaged in the transformation of information (i.e., data processing), services engaged in the provision of knowledge (i.e., design and engineering), and services aimed at the transformation of people (i.e., care of elderly people).

To be more precise, it is interesting to come up with the definition of service innovation. Consequently, it is worth referring to the work of Tether and Metcalfe [2004, p. 494] who defined a service innovation as "a new service experience or service solution that consists of one or several of the following dimensions: new service concept, new customer interaction, new value system/business partners, new revenue model, or new organizational or technological service delivery system." The six dimensions of service innovation are described in Table 8.2.

The OECD and Eurostat [2005, p. 38] argue about the diversity of the services sector, referring to the classification of the service sector by Howells *et al.* [2004] into four groups: services dealing mainly with

Table 8.2. Dimensions on Service Innovation.

Dimension	Explanation	Examples
Service concept	According to Den Hertog *et al.* [2010], the first dimension is referred to as the "service concept" or "service offering" and describes the value created by the service provider collaborating with the customer. The authors, citing Heskett [1986] and Frei [2008] argue that many new service concepts are a combination of elements of services that already exist and that are being incorporated in a new combination or configuration.	Telecom providers, for instance, integrating telephone, broadband, and TV into a single service concept.
New customer interaction	The second dimension is focused on the role that customers play in the creation of value. "The interaction process between the provider and the client is an important source of innovation — more so when the business service itself is offering support for innovation (which, for example, is the case in research and development (R&D) or design services)" [Den Hertog *et al.*, 2010, p. 494].	One valid example would be the different generations of electronic banking, ranging from the introduction of ATMs to online banking.
New value systems/new business partners	Refers to the involvement of actors to jointly produce service innovations. New services are increasingly realized through a combination of providers, different parties in the value chain, and actors in the value network. The literature has approached this concept in Chesbrough [2003] and Tee and Gawer [2009].	One example can be the combination of physical products and services, such as the iPhone and the iStore.

(*Continued*)

Table 8.2. (*Continued*)

Dimension	Explanation	Examples
New revenue models	A way to innovate in services is to focus on new revenue systems, which means to "find models to distribute costs and revenues in appropriate ways" [Den Hertog *et al.*, 2010, p. 495].	Some examples are switching from hiring specialists to building and operating transfer contracts in technical engineering, or moving from selling packaged software to ASP models in ICT.
New delivery system: personnel, organization, and culture	Originally coined by Heskett [1986], the "delivery system" refers to the organizational structure of the service itself. New services may require new organizational structures, interpersonal capabilities, or team skills.	Examples referred to the establishment of summer exchange programs [Normann, 1991] or the case of IKEA as not only an innovative retail concept but the way it is organized, empowers employees, or interacts with customers [Edvardsson and Enquist, 2011].
New delivery system: technological	Refers to new ICT-based innovation services ranging from e-government, e-health, customization of services, and virtual project teams.	A good example comes from the hospitality industry with online booking services.

Source: Den Hertog *et al.* [2010].

goods (such as transport and logistics), services dealing with information (such as call centers), knowledge-based services (such as design and engineering), and services dealing with people (such as healthcare). The *Oslo Manual* also pinpoints the blurred distinction between products and processes in the service sector, since production and consumption occur simultaneously. Thus, it states, "Services are intangible activities that are produced and consumed simultaneously and that change the conditions (e.g., physical, psychological, etc.) of users. The engagement of users through their time, availability, attention, transmission of information, or effort is often a necessary condition that leads to the co-production of services by users and the firm. The attributes or experience of a service can therefore, depend on the input of users" [OECD and Eurostat, 2018, p. 71].

As in the development of new products (goods) that we have studied earlier, success in new service development (NSD) doesn't occur by chance. Rather, it happens as a systematic and orchestrated process. Diverse investigations have stressed the importance of different elements that influence the success of NSD itself; i.e., Howells *et al.* [2004] found that frontline employees contribute to service innovation, and Zomerdijk and Voss [2011] analyzed NSD in experimental services finding that NSD practices such as multiple performance measures, cross-functional teams, and frontline involvement are more broadly used in this kind of service. This analysis also highlights the importance of service process innovation and the continuous innovation process requiring more flexibility, iterative processes, and non-linear processes. Kuester *et al.* [2013] developed a cluster analysis of 1,016 service companies, differentiating between efficient developers (transportation services, postal and telecommunications services, bank services, insurance services, sewage and refuse disposal, sanitation and similar activities); innovative developers (other services, data processing, research and development, consulting, construction services, technical, physical, and chemical analyses, other business services); interactive adopters (other financial services, wholesaling advertising, public administration, and defense); and standardized adopters (recreational, cultural and sporting services, retailing, real estate, renting, business activities, and domestic services). They also found 17 different NSD success determinants, which are displayed in Table 8.3.

It is also interesting to refer to a high level of coordination, qualification, and motivation among project team members as a way to increase success rates in NSD [Zomerdijk and Voss, 2011]; process formalization and the use of cross-functional and multidisciplinary teams that promote creativity, impact positively, and contribute to NSD speed [Froehle *et al.*, 2000]; and organizational factors such as the creation of an innovative climate and the combination and use of creativity-building techniques as ways to increase NSD success [Buganza and Verganti, 2006].

De Jong and Vermeulen [2003] made a literature review about the key factors that positively influenced the success of NSD. They found two stages: stage one refers to the management of key activities and stage two

Table 8.3. Success Determinants of NSD.

Success determinants of NSD
Service-related success factor categories
Service superiority
Service quality
Service newness
Quality of service experience
Process-related success factor categories
Market launch activities
Efficiency of service development process
Formal service development process
Customer integration
Company-related success factor categories
Staff competence
Synergy potential
Customer orientation
Internal cooperation
Top management support
Interdisciplinary teams
Service responsiveness
Innovation culture
Market-related success factor category
Market attractiveness

Source: Kuester *et al.* [2013, p. 535].

to the creation of an innovative climate. Concerning stage one, they referred to people-related activities such as the involvement of frontline employees, the presence of product champions and management support, and structure with activities such as funnel tools, multi-functional teams, availability of resources, pre-launch testing, and market research and launch. Concerning stage two, they included both people-related activities (external contacts, sharing information, and autonomy of employees) and structure-related activities (strategic focus, training and education, internal organizations and task rotation, and information technology).

Finally, Rapaccini *et al.* [2013] proposed a model for assessing the maturity of NSD processes in manufacturing companies that offer product services. The model entails five stages structured in the following dimensions: the approach used to manage processes and projects; the use of specific resources, skills, and tools; the involvement of customers, suppliers, and other stakeholders; and the adoption of performance management systems.

REFERENCES

Abernathy, W. J., and Utterback, J. M. [1978]. Patterns of industrial innovation, *Technology Review*, 80(7), pp. 40–47.

Barczak, G., Griffin, A., and Kahn, K. B. [2009]. Perspective: Trends and drivers of success in NPD practices: Results of the 2003 PDMA best practices study, *Journal of Product Innovation Management*, 26(1), pp. 3–23.

Baumann, H., Boons, F., and Bragd, A. [2002]. Mapping the green product development field: Engineering, policy and business perspectives, *Journal of Cleaner Production*, 10(5), pp. 409–425.

Bonner, J. M. [2010]. Customer interactivity and new product performance: Moderating effects of product newness and product embeddedness, *Industrial Marketing Management*, 39(3), pp. 485–492.

Booz, Allen, and Hamilton [1982]. *New Products Management for the 1980s* (Booz, Allen & Hamilton, New York).

Buganza, T., and Verganti, R. [2006]. Life-cycle flexibility: How to measure and improve the innovative capability in turbulent environments, *Journal of Product Innovation Management*, 23(5), pp. 393–407.

Capon, N., Farley, J. U., Lehmann, D. R., and Hulbert, J. M. [1992]. Profiles of product innovators among large US manufacturers, *Management Science*, 38(2), pp. 157–169.

Carbonell, P., Rodríguez-Escudero, A. I., and Pujari, D. [2009]. Customer involvement in new service development: An examination of antecedents and outcomes, *Journal of Product Innovation Management*, 26(5), pp. 536–550.

Chang, W., and Taylor, S. A. [2016]. The effectiveness of customer participation in new product development: A meta-analysis, *Journal of Marketing*, 80(1), pp. 47–64.

Chatterji, A. K., and Fabrizio, K. R. [2014]. Using users: When does external knowledge enhance corporate product innovation? *Strategic Management Journal*, 35(10), pp. 1427–1445.

Chesbrough, H. [2003]. The era of open innovation, *MIT Sloan Management Review*, 44(3), pp. 35–41.

Coombs, R., and Miles, I. [2000]. *Innovation, Measurement and Services: The New Problematique*, Metcalfe, J. S., and Miles, I. (eds.), *Innovation Systems in the Service Economy* (Springer, Boston) pp. 85–103.

Cooper, R. G. [1990]. Stage-gate systems: A new tool for managing new products, *Business Horizons*, 33(3), pp. 44–54.

Cooper, R. G. [1994]. Perspective third-generation new product processes, *Journal of Product Innovation Management*, 11(1), pp. 3–14.

Cooper, R. G. [2011]. *Winning at New Products: Creating Value Through Innovation*. 4th edn. (Basic Books, New York).

Cooper, R. G. [2013]. New products: What separates the winners from the losers and what drives success. P*DMA Handbook of New Product Development*, 3rd (edn.), (John Wiley & Sons, New Jersey) pp. 3–34.

Cooper, R. G., and Edgett, S. J. [2003]. Overcoming the crunch in resources for new product development, *Research-Technology Management*, 46(3), pp. 48–58.

Cooper, R. G., and Edgett, S. J. [2006]. Ten ways to make better portfolio and project selection decisions, P*DMA Visions Magazine*, 30(3), pp. 11–15.

Cooper, R. G., and Edgett, S. J. [2008]. Maximizing productivity in product innovation, *Research-Technology Management*, 51(2), pp. 47–58.

Cooper, R. G., and Edgett, S. J. [2010]. Developing a product innovation and technology strategy for your business, *Research-Technology Management*, 53(3), pp. 33–40.

Cooper, R. G., Edgett, S. J., and Kleinschmidt, E. J. [2002]. *Improving New Product Development Performance and Practices* (American Productivity & Quality Center, USA).

Cooper, R. G., Edgett, S. J., and Kleinschmidt, E. J. (2004a). Benchmarking best NPD practices I, *Research-Technology Management*, 47(1), pp. 31–43.

Cooper, R. G., Edgett, S. J., and Kleinschmidt, E. J. (2004b). Benchmarking best NPD practices II, *Research-Technology Management*, 47(3), pp. 50–59.

Cooper, R. G., Edgett, S. J., and Kleinschmidt, E. J. (2004c). Benchmarking best NPD practices III, *Research-Technology Management*, 47(6), pp. 43–55.

Cooper, R. G., Edgett, S. J., and Kleinschmidt, E. J. [2006]. *Portfolio Management for New Product Development*, Product Development Institute, USA.

Cooper, R. G., and Kleinschmidt, E. J. [1995]. Benchmarking the firm's critical success factors in new product development, *Journal of Product Innovation Management*, 12(5), pp. 374–391.

Cooper, R. G., and Kleinschmidt, E. J. [2007]. Winning businesses in product development: The critical success factors, *Research-Technology Management*, 50(3), pp. 52–66.

Coviello, N. E., and Joseph, R. M. [2012]. Creating major innovations with customers: Insights from small and young technology firms, *Journal of Marketing*, 76(6), pp. 87–104.

Crawford, C. M. [1987]. New product failure rates: A reprise, *Research Management*, 30(4), pp. 20–24.

De Jong, J. P., and Vermeulen, P. A. [2003]. Organizing successful new service development: A literature review, *Management Decision*, 41(9), pp. 844–858.

Den Hertog, P., Van der Aa, W., and De Jong, M. W. [2010]. Capabilities for managing service innovation: Towards a conceptual framework, *Journal of Service Management*, 21(4), pp. 490–514.

Djellal, F., and Gallouj, F. [2001]. Patterns of innovation organization in service firms: Postal survey results and theoretical models, *Science and Public Policy*, 28(1), pp. 57–67.

Dooley, K. J., Subra, A., and Anderson, J. [2002]. Adoption rates and patterns of best practices in new product development, *International Journal of Innovation Management*, 6(01), pp. 85–103.

Drejer, I. [2004]. Identifying innovation in surveys of services: A Schumpeterian perspective, *Research Policy*, 33(3), pp. 551–562.

Edgett, S. J. [2011]. *New Product Development: Process Benchmarks and Performance Metrics*, Stage-Gate International, USA.

Edvardsson, B., and Enquist, B. [2011]. The service excellence and innovation model: Lessons from IKEA and other service frontiers, *Total Quality Management & Business Excellence*, 22(5), pp. 535–551.

Eiglier, P., and Langeard, E. [1977]. A new approach to service marketing, Eiglier, P., Langeard, E., Lovelock, C. H., Bateson, J. E. G., and Young, R. F. (eds.), *Marketing Consumer Services: New Insights*, Report No.77–115, Marketing Science Institute, Cambridge.

Enkel, E., Gassmann, O., and Chesbrough, H. [2009]. Open R&D and open innovation: Exploring the phenomenon, *R&D Management*, 39(4), pp. 311–316.

Fang, E. [2008]. Customer participation and the trade-off between new product innovativeness and speed to market, *Journal of Marketing*, 72(4), pp. 90–104.

Fang, E., Palmatier, R. W., and Evans, K. R. [2008]. Influence of customer participation on creating and sharing of new product value, *Journal of the Academy of Marketing Science*, 36(3), pp. 322–336.

Filippini, R., Salmaso, L., and Tessarolo, P. [2004]. Product development time performance: Investigating the effect of interactions between drivers, *Journal of Product Innovation Management*, 21(3), pp. 199–214.

Frei, F. X. [2008]. The four things a service business must get right, *Harvard Business Review*, 86(4), pp. 70–80.

Froehle, C. M., Roth, A. V., Chase, R. B., and Voss, C. A. [2000]. Antecedents of new service development effectiveness: An exploratory examination of strategic operations choices, *Journal of Service Research*, 3(1), pp. 3–17.

Fuchs, C., and Schreier, M. [2011]. Customer empowerment in new product development, *Journal of Product Innovation Management*, 28(1), pp. 17–32.

Gallouj F. [2000]. *Beyond technological innovation: Trajectories and varieties of service innovations*, Boden M., and Miles I. (eds.), *Services and the Knowledge based Economy* (Continuum International Publishing Group, London) pp. 129–145.

Gallouj, F., and Weinstein, O. [1997]. Innovation in services, *Research Policy*, 26(4–5), pp. 537–556.

Gmelin, H., and Seuring, S. [2014]. Determinants of a sustainable new product development, *Journal of Cleaner Production*, 69, pp. 1–9.

Grönroos, C. [2007]. *Service Management and Marketing: Customer Management in Service Competition* (John Wiley & Sons, Hoboken, New Jersey, USA).

Hagedoorn, J. [2002]. Inter-firm R&D partnerships: An overview of major trends and patterns since 1960, *Research Policy*, 31(4), pp. 477–492.

Hauknes, J., and Miles, I. [1996]. *Services in European Innovation Systems — A Review of Issues* (STEP, Oslo).

Heskett, J. L. [1986]. *Managing in the Service Economy* (Harvard Business Press, Boston).

Hilletofth, P., and Eriksson, D. [2011]. Coordinating new product development with supply chain management, *Industrial Management & Data Systems*, 111(2), pp. 264–281.

Hong, P., Doll, W. J., Revilla, E., and Nahm, A. Y. [2011]. Knowledge sharing and strategic fit in integrated product development projects: An empirical study, *International Journal of Production Economics*, 132(2), pp. 186–196.

Howells, J., Tether, B., Gallouj, F., Djellal, F., Gallouj, C., Blind, K., Corrocher, N. [2004]. *Innovation in Services: Issues at Stake and Trends* (European Commission, Bruxelles).

Hullova, D., Trott, P., and Simms, C. D. [2016]. Uncovering the reciprocal complementarity between product and process innovation, *Research Policy*, 45(5), pp. 929–940.

Iansiti, M., and MacCormack, A. [1997]. Developing products on Internet time, *Harvard Business Review*, 75(5), pp. 108–117.

Johnsen, T. E. [2009]. Supplier involvement in new product development and innovation: Taking stock and looking to the future, *Journal of Purchasing and Supply Management*, 15(3), pp. 187–197.

Joshi, A. W., and Sharma, S. [2004]. Customer knowledge development: Antecedents and impact on new product performance, *Journal of Marketing*, 68(4), pp. 47–59.

Kahn, K. B. [2012]. *The PDMA Handbook of New Product Development*, 3rd edn. (Wiley, Hoboken)

Kahn, K. B., Barczak, G., and Moss, R. [2006]. Perspective: Establishing an NPD best practices framework, *Journal of Product Innovation Management*, 23(2), pp. 106–116.

Kahn, K. B., Barczak, G., Nicholas, J., Ledwith, A., and Perks, H. [2012]. An examination of new product development best practice, *Journal of Product Innovation Management*, 29(2), pp. 180–192.

Kessler, E. H., Bierly, P. E., and Gopalakrishnan, S. [2000]. Internal vs. external learning in new product development: Effects on speed, costs and competitive advantage, R&D *Management*, 30(3), pp. 213–224.

Kleinschmidt, E. J., and Cooper, R. G. [1991]. The impact of product innovativeness on performance, *Journal of Product Innovation Management*, 8(4), pp. 240–251.

Klette, T. J., and Griliches, Z. [2000]. Empirical patterns of firm growth and R&D investment: A quality ladder model interpretation, *The Economic Journal*, 110(463), pp. 363–387.

Klette, T. J., and Kortum, S. [2004]. Innovating firms and aggregate innovation, *Journal of Political Economy*, 112(5), pp. 986–1018.

Kotler, P., Wong, V., Saunders, J., and Armstrong, G. [2002]. *Principles of Marketing*, 3rd edn. (Pearson Education Limited, Essex).

Koufteros, X. A., Rawski, G. E., and Rupak, R. [2010]. Organizational integration for product development: The effects on glitches, on-time execution of engineering change orders, and market success, *Decision Sciences*, 41(1), pp. 49–80.

Koufteros, X., Vonderembse, M., and Jayaram, J. [2005]. Internal and external integration for product development: The contingency effects of uncertainty, equivocality, and platform strategy, *Decision Sciences*, 36(1), pp. 97–133.

Kuester, S., Schuhmacher, M. C., Gast, B., and Worgul, A. [2013]. Sectoral heterogeneity in new service development: An exploratory study of service types and success factors, *Journal of Product Innovation Management*, 30(3), pp. 533–544.

Lager, T. [2002]. A structural analysis of process development in process industry: A new classification system for strategic project selection and portfolio balancing, *R&D Management*, 32(1), pp. 87–95.

Langerak, F., Hultink, E. J., and Robben, H. S. [2004]. The impact of market orientation, product advantage, and launch proficiency on new product performance and organizational performance, *Journal of Product Innovation Management*, 21(2), pp. 79–94.

Leiponen, A. [2002]. Intellectual property and innovation in business services: Implications for the management of knowledge and supply relationships, *DRUID 2002 Summer Conference on Industrial Dynamics of the New and Old Economy — Who is Embracing Whom*, Druid Society, pp. 6–8.

Loch, C., Stein, L., and Terwiesch, C. [1996]. Measuring development performance in the electronics industry, *Journal of Product Innovation Management*, 13(1), pp. 3–20.

Lynn, G. S., Skov, R. B., and Abel, K. D. [1999]. Practices that support team learning and their impact on speed to market and new product success, *Journal of Product Innovation Management*, 16(5), pp. 439–454.

Mahajan, V., and Wind, Y. [1988]. New product forecasting models: Directions for research and implementation, *International Journal of Forecasting*, 4(3), pp. 341–358.

Mahr, D., Lievens, A., and Blazevic, V. [2014]. The value of customer cocreated knowledge during the innovation process, *Journal of Product Innovation Management*, 31(3), pp. 599–615.

Marion, T. J., Friar, J. H., and Simpson, T. W. [2012]. New product development practices and early-stage firms: Two in-depth case studies, *Journal of Product Innovation Management*, 29(4), pp. 639–654.

McNally, R. C., Akdeniz, M. B., and Calantone, R. J. [2011]. New product development processes and new product profitability: Exploring the mediating role of speed to market and product quality, *Journal of Product Innovation Management*, 28(s1), pp. 63–77.

Menon, A., Chowdhury, J., and Lukas, B. A. [2002]. Antecedents and outcomes of new product development speed: An interdisciplinary conceptual framework, *Industrial Marketing Management*, 31(4), pp. 317–328.

Miles, I. [2005]. Knowledge intensive business services: Prospects and policies, *Foresight*, 7(6), pp. 39–63.

Milgrom, P., and Roberts, J. [1995]. Complementarities and fit strategy, structure, and organizational change in manufacturing, *Journal of Accounting and Economics*, 19(2–3), pp. 179–208.

Monczka, R. M., Handfield, R. B., Scannell, T. V., Ragatz, G. L., and Frayer, D. J. [2000]. *New Product Development-Strategies for Supplier Involvement* (ASQ Quality Press, Milwaukee, V).

Morgan, T., Obal, M., and Anokhin, S. [2018]. Customer participation and new product performance: Towards the understanding of the mechanisms and key contingencies, *Research Policy*, 47(2), pp. 498–510.

Nakata, C., and Im, S. [2010]. Spurring cross-functional integration for higher new product performance: A group effectiveness perspective, *Journal of Product Innovation Management*, 27(4), pp. 554–571.

Ngo, L. V., and O'cass, A. [2013]. Innovation and business success: The mediating role of customer participation, *Journal of Business Research*, 66(8), pp. 1134–1142.

Normann, R. [1991]. *Service Management: Strategy and Leadership in Service Business* (Wiley, London).

OECD and Eurostat [2005]. O*slo Manual* (OECD Publishing, France).

OECD and Eurostat [2018]. O*slo Manual* (OECD Publishing, France).

Page, A. L. [1993]. Assessing new product development practices and performance: Establishing crucial norms, *Journal of Product Innovation Management*, 10(4), pp. 273–290.

Perrone, G., Roma, P., and Nigro, G. L. [2010]. Designing multi-attribute auctions for engineering services procurement in new product development in the automotive context, *International Journal of Production Economics*, 124(1), pp. 20–31.

Petersen, K. J., Handfield, R. B., and Ragatz, G. L. [2003]. A model of supplier integration into new product development, *Journal of Product Innovation Management*, 20(4), pp. 284–299.

Pisano, G. P. [1997]. *The Development Factory: Unlocking the Potential of Process Innovation* (Harvard Business Press, Boston).

Pisano, G. P., and Wheelwright, S. C. [1995]. The new logic of high-tech R & D, *Long Range Planning*, *Harvard Business Review*, 73(5), pp. 93–105.

Ragatz, G. L., Handfield, R. B., and Petersen, K. J. [2002]. Benefits associated with supplier integration into new product development under conditions of technology uncertainty, *Journal of Business Research*, 55(5), pp. 389–400.

Rapaccini, M., Saccani, N., Pezzotta, G., Burger, T., and Ganz, W. [2013]. Service development in product-service systems: A maturity model, *The Service Industries Journal*, 33(3–4), pp. 300–319.

Song, M., and Di Benedetto, C. A. [2008]. Supplier's involvement and success of radical new product development in new ventures, *Journal of Operations Management*, 26(1), pp. 1–22.

Tang, C. S. [2006]. Perspectives in supply chain risk management, *International Journal of Production Economics*, 103(2), pp. 451–488.

Tee, R., and Gawer, A. [2009]. Industry architecture as a determinant of successful platform strategies: A case study of the i-mode mobile Internet service, *European Management Review*, 6(4), pp. 217–232.

Tether, B. S., and Metcalfe, J. S. [2004]. *Systems of Innovation in Services. Sectoral Systems of Innovation and Production in Europe* (Cambridge University Press, Cambridge).

Townsend, J. D., Cavusgil, S. T., and Baba, M. L. [2010]. Global integration of brands and new product development at general motors, *Journal of Product Innovation Management*, 27(1), pp. 49–65.

Trott, P. [2008]. *Innovation Management and New Product Development* (Pearson Education, USA).

Troy, L. C., Hirunyawipada, T., and Paswan, A. K. [2008]. Cross-functional integration and new product success: An empirical investigation of the findings, *Journal of Marketing*, 72(6), pp. 132–146.

Un, C. A., Cuervo-Cazurra, A., and Asakawa, K. [2010]. R&D collaborations and product innovation, *Journal of Product Innovation Management*, 27(5), pp. 673–689.

Utterback, J., Vedin, B.-A., Alvarez, E., Ekman, S., Walsh Sanderson, S., Tether, B., and Verganti, R. [2006]. Design-inspired innovation and the design discourse, *Design-Inspired Innovation*, pp. 154–186.

van Echtelt, F. E., Wynstra, F., and van Weele, A. J. [2007]. Strategic and operational management of supplier involvement in new product development: A contingency perspective, *IEEE Transactions on Engineering Management*, 54(4), pp. 644–661.

Vargo, S. L., and Lusch, R. F. [2004]. The four service marketing myths: Remnants of a goods-based, manufacturing model, *Journal of Service Research*, 6(4), pp. 324–335.

von Corswant, F., and Tunälv, C. [2002]. Coordinating customers and proactive suppliers: A case study of supplier collaboration in product development, *Journal of Engineering and Technology Management*, 19(3), pp. 249–261.

von Hippel, E. [1976]. The dominant role of users in the scientific instrument innovation process, *Research Policy*, 5(3), pp. 212–239.

von Hippel, E. [1986]. Lead users: A source of novel product concepts, *Management Science*, 32(7), pp. 791–805.

Wagner, S. M. [2010]. Supplier traits for better customer firm innovation performance, *Industrial Marketing Management*, 39(7), pp. 1139–1149.

Wagner, S. M. [2012]. Tapping supplier innovation, *Journal of Supply Chain Management*, 48(2), pp. 37–52.

Yli-Renko, H., and Janakiraman, R. [2008]. How customer portfolio affects new product development in technology-based entrepreneurial firms, *Journal of Marketing*, 72(5), pp. 131–148.

Zhang, Q., and Doll, W. J. [2001]. The fuzzy front end and success of new product development: A causal model, *European Journal of Innovation Management*, 4(2), pp. 95–112.

Zomerdijk, L. G., and Voss, C. A. [2011]. NSD processes and practices in experiential services, *Journal of Product Innovation Management*, 28(1), pp. 63–80.

CHAPTER 9

EFFECTS OF INNOVATION ON THE FIRM PERFORMANCE

9.1 THE OBJECTIVES OF INNOVATION

The general hypothesis about innovation is that it improves business results and firm competitiveness. Different models and approaches to innovation have depicted a process through which science, technology, business processes, and the market interact [Rothwell, 1994]. Thus, the challenge is to determine to what extent the investments made in innovation are returned to the market and what are their financial results. To achieve this, the first point is examining the business objectives for innovation. Accordingly, Dosi [1988, p. 1120] argues that agents in the market are profit-seekers and they allocate resources to the development of new products and new techniques of production if they think they fulfill three conditions: "Some sort of yet-unexploited scientific and technical opportunities; if they expect that there will be a market for their new products and processes; and, finally, if they expect some economic benefit, net of the incurring costs, deriving from the innovations." Therefore, the author underlines the economic character of innovation that is supported by the existence of a technical or scientific opportunity, the belief that there is a demand for a new kind of goods and services, and the possibility of obtaining economic benefits. The latter is quite relevant to the object of our analysis, since it encourages the economic orientation of innovation.

In the same vein, the *Oslo Manual* in its 4th edition refers to the fact that both individuals and organizations seek a clear benefit from innovation. Accordingly, the cited manual posits, "decisions to innovate can be presumed, *a priori*, to have an implicit motive to directly or indirectly benefit the innovative organization, community or individual," and states that, "in the business enterprise sector, benefits often involve profitability. In normally functioning markets, customers have the freedom to decide whether to acquire a new product on the basis of its price and characteristics. Therefore, the markets for products and finance fulfill a selection function for innovations by guiding the processes of resource allocation in the business enterprise sector" [OECD, 2018, p. 48]. Likewise, the 3rd edition of the *Oslo Manual* [OECD, 2005, p. 34] clearly refers to the economic character of business innovation objectives making reference directly to business performance, and postulates, "innovation in firms refers to planned changes in a firm's activities with a view to improving the firm's performance." In particular, it is believed that innovation can improve a firm's performance by different means [OECD, 2005, p. 35] as follows:

- gaining a competitive advantage (or simply maintaining competitiveness);
- shifting the demand curve of the firm's products (e.g., increasing product quality, offering new products, or opening up new markets or groups of customers);
- shifting the firm's cost curve (e.g., reducing unit costs of production, purchasing, distribution, or transactions);
- improving the firm's ability to innovate (e.g., increasing the ability to develop new products or processes or to gain and create new knowledge).

In the same regard, Gopalakrishnan and Damanpour [1997, p. 20] point out the importance of innovation as a means to increase productivity at both macro- and microlevels, and postulate, "economists view innovation as a phenomenon that both brings about large changes in productivity at the industry and firm level and explains inter-industry variability in growth, productivity and overall performance." Likewise, Bettis and Hit [1995, p. 14] draw attention to the changing nature of the market and stress the need for innovation irrespective of the degree of novelty, and argue, "because of the dynamism of the new competitive landscape, firms cannot remain static even if they operate in mature industries. Incremental (and perhaps even radical) innovation may lengthen the product life cycle and change the competitive dynamics within the market."

9.2 DEFINING BUSINESS PERFORMANCE

Keeping in mind the aforementioned discussion, we assume that innovation aims to achieve a competitive advantage for the firm that will result in better business performance. Therefore, to delve into the relationship between innovation and firm performance, we first have to properly outline what business performance is and how it can be measured. In general, organizational performance measurement has been addressed in the literature on innovation through the creation of different constructs [Venkatraman and Ramanujam, 1986]. For instance, Hogan and Coote [2014, p. 1615] distinguish between market and financial performance constructs made up of different variables and define market performance as the "degree to which an organization attracts and retains customers for its products and services", while financial performance is defined as "the degree to which an organization achieves economic outcomes." The authors established a set of indicators or variables to measure both market and financial performance, as depicted in Table 9.1.

In the same regard, Murphy *et al.* (1996) classified performance measures as financial performance and operational performance. The authors conducted an analysis of the most common performance measures used in the scientific literature and then characterized eight different dimensions of performance: efficiency, growth, profit, size, liquidity, success and failure, market share, and leverage.

To delve into the meaning of firm performance, it is also worth citing the work of Hagedoorn and Cloodt [2003], who distinguish between the concepts of inventive performance, technological performance, and innovative performance. Table 9.2 contains the definitions of each concept.

Table 9.1. Market and Financial Performance Indicators.

Market performance	Financial performance
1. Achieving client satisfaction	1. Overall profitability
2. Providing value for clients	2. Profitability per employee
3. Keeping current clients	3. Profit growth
4. Attracting new clients	4. Overall cash flow
5. Attaining desired growth	5. Cash flow per employee
6. Securing desired market share	6. Growth in cash flow

Source: Hogan and Coote [2014, p. 1619].

Table 9.2. Definitions of Inventive, Technological, and Innovative Performance.

Concept	Definition
Inventive performance	"The achievements of companies in terms of ideas, sketches, models of new devices, products, processes and systems."
Technological performance	"The accomplishment of companies with regard to the combination of their R&D input, as an indicator of their research capabilities, and their R&D output in terms of patents."
Innovative performance (narrow sense)	"Results for companies in terms of the degree to which they actually introduce inventions into the market; i.e., their rate of introduction of new products, new process systems or new devices."
Innovative performance (broad sense)	"Focuses on both the technical aspects of innovation and the introduction of new products into the market, but it excludes the possible economic success of innovations as such."

Source: Hagedoorn and Cloodt [2003, p. 1367].

This differentiation is relevant, since it helps to visualize the various outputs depending on the phase or stage of the innovation process. Therefore, we have to consider that the measurement of performance can be applied to any segment of the innovation continuum, starting from the generation of new knowledge to the launching of new products, the implementation of new processes, and lastly, the generation of market or financial returns. However, it seems that the categorization and measurement of performance it is not an easy task, as has been noted in the innovation literature over time [Hagedoorn and Cloodt, 2003]. These authors refer to the availability of general measures "such as R&D inputs, patent counts, patent citations, or counts of new product announcements, and more specific survey-based measurements of this particular performance by companies have been used in trying to capture this innovative performance of companies" [Hagedoorn and Cloodt, 2003, p. 1365].

To shed light on the relationship between the innovation process and its performance, it is worth citing the work of Basberg [1987], which describes the relationship between inventions, innovations, and patents. The author argues that only a portion of the knowledge generated by the firm through R&D processes will result in inventions. Then, from the whole inventive stock of the firm, only a part will be protected by patents. Finally, some of the inventions and patents will get to the market through

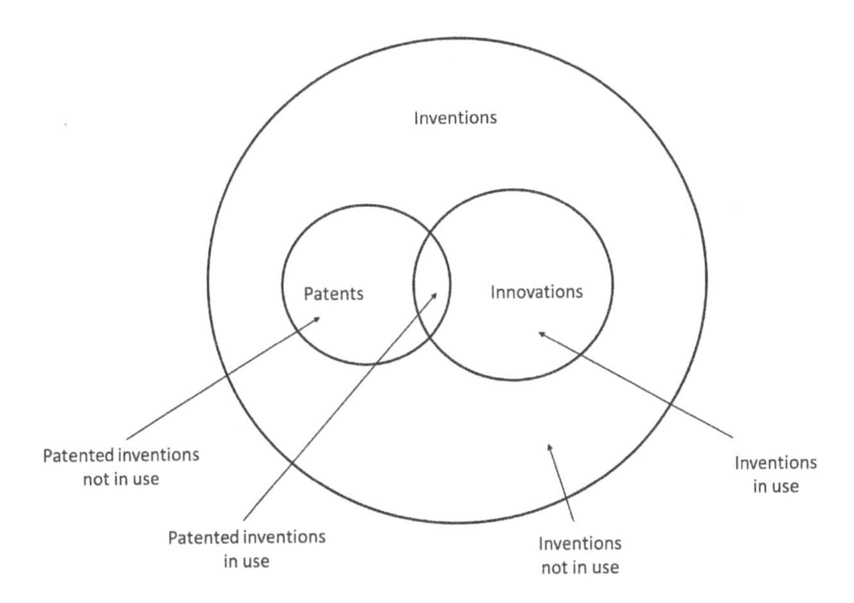

Figure 9.1. Relationship Between Inventions, Patents, and Innovations [Basberg, 1987, p. 133].

the implementation of new processes and products, while others will remain useless. Furthermore, we can also hypothesize that new products and processes will result in better performance of the firm. This is represented in Fig. 9.1, where the big circle depicts the R&D effort that has materialized in inventions. Likewise, both inside circles — patents and innovations — represent those inventions that have been protected (patents) and those that have become innovations. The intersection between patents and innovations is especially significant, since it represents the patent-based innovations.

9.3 RELATIONSHIP BETWEEN INNOVATION AND FIRM PERFORMANCE

Specifically, the intent of this chapter is to delve into the impact of innovation on firm performance from the financial or market points of view. Given the seminal conceptualization of innovation, economists correlated early adoption of technology and innovations to profitability and superior performance of the firm. Schumpeter [1934] argues that when new ideas enter the market, firms that are stuck to the old ones risk destruction, expecting businesses to promptly adopt innovation in order to gain competitiveness. Accordingly, innovative products that are first introduced into the market face little competition and are able to achieve transitory monopolies that are ultimately ended by new market entrants, which ultimately reduces margins and profits. Therefore, the only way to maintain a high level of profit is to introduce a continuous stream of new products into the market to disrupt reduced margins and profits in a process of "creative destruction" [Schumpeter, 1942].

Therefore, according to the scientific literature, innovation seems to be one of the key elements in guaranteeing the competitiveness of the firm. For instance, Jiménez and Sanz-Valle [2011] conducted a study with data from 451 Spanish firms and found that organizational learning and innovation contribute positively to business performance. In the same vein, Gunday *et al.* [2011] argue that innovation has an expected impact on production and the market itself. The authors postulate that "innovations can be expected to lead to improvements in production and market performances through the mediation of innovative performance. In this respect, innovative performance plays the role of an effective hub that carries the positive effects of innovations to the various aspects of firm performance" [Gunday, *et al.*, 2011, p. 665]. Consequently, the authors refer to "innovativeness" as an intermediate output of the business strategy that impacts firm performance.

To strengthen the analysis, we cite Lumpkin and Dess [1996, p. 142], who offer a definition of innovativeness as an organization's "tendency to engage in and support new ideas, novelty, experimentation, and creative processes that may result in new products, services, or technological processes." Also, García and Calantone [2002, p. 113] define innovativeness as "the capacity of a new innovation to influence the firm's existing marketing resources, technological resources, skills, knowledge, capabilities, or strategy." Similarly, some studies investigate the relationship between innovativeness and firm performance; for instance, Keskin [2006, p. 411] argues that firm innovativeness positively affects firm performance and posits that "learning-orientation translates marketing attitudes into effective behavior to facilitate innovation" as a means to encourage innovativeness in the firm. Accordingly, the author details that a firm's market orientation indirectly impacts the firm performance via the firm innovativeness and learning.

Roberts [1999] also conducted an analysis of the U.S. pharmaceutical industry, finding a positive relationship between innovative propensity and sustained superior profitability. The authors argue that innovation helps safeguard the overall high-performance position of the firm over time. In the same regard, Gunday *et al.* [2011, p. 665] claim that "innovative performance can exert then positive effects on firms' production, market and financial performances in the long-term." However, the same authors citing Lawless and Anderson [1996] argue that the adoption of technologies involves an initial penalty and it needs a period of time to demonstrate positive impacts on the firm performance [Damanpour and Evan, 1984].

Similarly, Cho and Pucik [2005] studied the U.S. finance industry at the firm level. They investigated the association between innovativeness and firm performance, which was measured through quality, growth, profitability, and market value. The authors determined that "quality alone is not sufficient to create high growth, and innovativeness alone is not sufficient to improve profitability" [Cho and Pucik, 2005, p. 573]. Therefore, it seems that quality leverages the effect of innovativeness and vice versa, reaching the conclusion that the overall strategy of the firm should balance both innovation and quality in a win-win approach.

Also, Calantone *et al.* [2002, p. 522] argue that firm innovativeness shows a positive correlation to firm performance. The authors state that competitive advantage is built on the understanding of customer needs, competition, and technological development. These authors developed a logic framework that relates learning orientation (commitment to learning, shared vision, open-mindedness, intra-organizational knowledge sharing) to firm innovativeness and performance. Consequently, the results qualify innovation as a broad process of learning that leads to the implementation of new ideas, products, and processes.

Similarly, Han *et al.* (1998) examined U.S. banking industry data and found that market orientation impacts organizational innovativeness as a mediating variable of corporate performance. The authors consider market orientation a construct of customer orientation, competitor orientation, and inter-functional coordination. Specifically, the authors underline the importance of innovation to firm performance and state, "we reaffirmed that innovations, as vital components of business performance, warrant organization-wide attention for successful implementation of both technical and administrative kinds" [Han *et al.*, 1998, p. 41].

In the same regard, Salavou [2002, p. 168] investigated the food industry in Greece and found evidence of a positive relationship between innovativeness and firm performance, measured in this case by the variable of business profitability. In particular, the authors discovered that market orientation, in terms of customer responsiveness, market-driven pricing policy, and product innovation interact in affecting business profitability. Specifically, the authors found that firms respond to customer needs offering less radical or incremental product innovations, whereas while offering more radical product innovations, the higher was their profitability in terms of return on assets (ROA). In the same regard, Weerawardena (2003) highlights the importance of market orientation as a source of innovation and competitive advantage. The author particularly stresses the relevance of market-focused learning capability and organizational innovation as a means to sustain competitive advantage. Similarly, Avlonitis and Salavou [2007] analyzed the role of entrepreneurial orientation (EO) in business performance. The authors found evidence of EO's impact on product innovativeness and product performance.

As well, Pelham [2000], investigating small and medium-sized manufacturing firms, found that market orientation has a strong relationship with firm performance, noticing that this link is stronger than strategy, firm size, or industry characteristics. Also, Wolff and Pett [2006] studied firm performance in small and medium-sized enterprises (SMEs) and demonstrated that product improvement orientation is positively associated with growth and, in turn, profitability; whereas the process improvement orientation showed no statistical relationship to growth and, ultimately, profitability. The authors considered environmental hostility, firm size, innovation capability, and internationalization to measure the impact on firm performance and established that innovativeness has a clear impact on the performance of the organization. Also, Terziovski [2010] investigated a sample of 600 SMEs in Australia and discovered that, equally to large businesses, innovation strategy and formal structure are key drivers of performance, pointing out that innovation culture and strategy are closely related in the innovation process.

Finally, Hanel and St. Pierre [2006, p. 496], based on data from the Statistics Canada Survey of Innovation, studied industry businesses to investigate the relationship between collaboration with universities and firm performance. Collaboration between university and industry can increase the firm's ability to recognize, absorb, and apply externally received knowledge, which is critical to their innovative capabilities [Kande *et al.*, 2017]. Therefore, collaboration is related to the innovativeness of the firm and can leverage other capabilities of the business. The study shows that the most profitable innovations introduced by firms are those resulting from collaboration with university partners, which led to superior economic performance compared to non-collaborating firms. Likewise, collaboration with universities is positively related to the originality of innovations and the economic performance of the innovating firm (competitive position, profit margins, share of international markets, and increased profitability).

Now that we have specifically analyzed the relationship between innovativeness and firm performance; we continue with an investigation of the effects of innovation on firm performance attending to different variables. To this end, we cite noteworthy research from Damanpour and Evan [1984], which examined the impact of the adoption of administrative and technical innovation on business performance. They applied the organizational lag model, finding that the pace of adoption of technical innovations is faster than administrative innovations. The authors studied the lag model in 85 public libraries to measure the impact of organizational types on organizational performance and found that the degree of organizational lag is inversely related to organizational performance.

Continuing with this analysis, it is also remarkable to cite the work of Thornhill [2006], which studied 845 Canadian manufacturing firms for industry dynamism. The author posits that though innovation is more common when industry dynamism is high, innovative firms are likely to enjoy revenue growth, irrespective of the kind of industry in which they operate. The author also argues that firm knowledge, industry dynamism, and innovation interact in the way they influence firm performance. The results imply that a highly skilled workforce is crucial to the firm performance in dynamic environments, while investments in training leverage result in stable environments. Shaukat *et al.* [2013] studied the relationship of innovation to business performance (innovativeness, market, production, and financial) with a sample of 150 manufacturing firms in Pakistan. They evidenced the link between innovativeness and firm performance, reporting a strong effect of process innovation on firm performance.

Another interesting area of study is the link between research and development (R&D) and firm performance. In particular, there is a huge body of contributions regarding innovation and R&D effects on firm growth. The research reports a positive relationship between firm growth and innovation that differs according to firm characteristics, the nature of market selection, and the geographical environment [Audretsch *et al.*, 2014]. For instance, García-Manjón and Romero-Merino [2012] presented a model of endogenous firm growth considering R&D investment as one of the main variables. The authors' investigation evidences a positive effect of R&D intensity on sales growth for a sample of 754 European firms for the 2003–2007 period. The authors also argue that the effect of R&D on sales growth is more intense in high-growth firms and is especially significant when referring to high-technology sectors. Also, Coad and Rao [2008] noted that the positive relationship and impact of innovative activities on firm growth are mainly focused on the fastest growing firms. In the same vein, Demirel and Mazzucato [2012] explored the link between innovation and firm growth for small and large publicly quoted U.S. pharmaceutical firms between 1950 and 2008. The authors found that firm characteristics (firm size, number of patents, or persistence in patenting) condition the effect of R&D on growth. For instance, R&D positively impacts growth for large patenting firms, while for small firms, R&D only triggers growth in those that persistently patent for a minimum of 5 years. Similarly, Calabrese *et al.* [2013], using data from high-tech industries (biotech and aerospace) in Italy, state that innovation efforts, innovation protection, and risk propensity have positive effects on the economic performance of such enterprises.

As we discussed earlier, the balance of the R&D input–output equation, usually measured in terms of patents, shows the technological performance of the firm and that it is an indicator of its research capabilities.

This capacity is a valuable resource for improving business innovation and firm performance and has been addressed throughout the scientific literature. In this vein, Artz *et al.* [2010, p. 733] analyzed a sample of 272 firms in 35 industries over 19 years to investigate the relationships between R&D and innovative output at the firm level. The main findings were that product announcements and patents were positively correlated. Likewise, R&D spending is positively correlated with both product announcements and patents. At the same time, product announcements were positively correlated with firm performance, measured by ROA and sales growth. Similarly, Griliches *et al.* [1986] stress the value of patents as indicators of innovation activity and point out the relationship between R&D expenditures, the level of patenting, and the stock market value of firms. Likewise, Narin *et al.* [1987] studied 17 U.S. pharmaceutical companies to examine the relationship between corporate patent and patent citation data, and several other indicators of corporate performance. The authors found strong correlations between increases in company profits and sales, and both patent citation frequency and concentration of company patents within a few patent classes. Similarly, Sohn *et al.* [2010] examined the impact of R&D and patents on the financial performance of venture firms in South Korea. The results confirm a positive relationship between R&D investment and patenting, even though there were no conclusive results showing the effect of patenting on financial performance of the firm. The study also confirms that R&D is positively correlated with sales growth, whereas R&D and patents did not improve firm profitability.

In the same regard, Ernst [2001] analyzed 50 German machine tool manufacturers in a panel analysis between 1984 and 1992 to study the relationship between patent-filing applications and firm performance. The author determined that patent applications were positively linked to firm performance. In particular, the results showed that national patent applications lead to sales increases within a span of 2–3 years, whereas European patent applications drove even higher sales after 3 years from the reference year. Moreover, Mann and Sager [2007] investigated software start-up firms and found significant positive correlations between patenting and some variables rating the firm's performance (including number of rounds, total investment, exit status, receipt of late-stage financing, and longevity). Finally, DeCarolis and Deeds [1999] used a knowledge-based view of the firm to depict the relationship between firm capabilities and firm performance. The authors consider R&D expenditures representative of knowledge flows, while products in the pipeline, firm citations, and patents were indicative of knowledge stocks. The results show that products in the pipeline and firm citations are predictors of firm performance.

Continuing with the analysis, the scientific literature evidences more innovation-related variables that influence business performance. For example, Therrien *et al.* [2011] cited the impact of innovation novelty on firm performance in service industries, finding that service firms that enter earlier in the market or introduce high-novelty products achieve better firm performance, in particular more sales. The results demonstrate that firm performance based on higher sales is influenced by early-entry or higher novelty levels. Accordingly, Lieberman and Montgomery [1988, p. 6] argue, "in many markets, there is room for only a limited number of profitable firms; the first mover can often select the most attractive niches and may be able to take strategic actions that limit the amount of space available for subsequent entrants." Also, in the service industry, Mansury and Love [2008] studied 206 U.S. business firms and evidenced that service innovation has a consistently positive effect on growth as a measure of firm performance. The authors refer to the importance of external linkages that the innovators maintain through the innovation process in order to explain the effect on growth. Particularly, variables such as customer orientation and the existence of joint venture partners were determined to be relevant in explaining firm growth.

Regarding industrial and production firms, Gunday *et al.* [2011, p. 27] conducted a study of 184 manufacturing firms in Turkey. The authors found support for the hypothesis that innovations performed in manufacturing firms have positive and significant impacts on innovative performance. Likewise, innovative firms have higher market share, total sales, and exports. Also in the industrial realm, Geroski *et al.* [1993] analyzed 721 manufacturing firms in the UK and found that innovation has a positive relationship with operating profit margin. They pointed out that, although the impact on profits was modest, innovative firms in general were more profitable than non-innovative firms. In particular, Atalay *et al.* [2013, p. 233] analyzed the relationship between innovation types and firm performance within the context of the Turkish automotive supplier industry. The results evidence that product and process innovation positively and significantly affect firm performance. Finally, Roper and Love [2002] studied the determinants of export performance in UK and German manufacturing plants. The results showed a positive correlation between product innovation and the probability and propensity to export.

Also interesting is the work of López-Nicolás and Meroño-Cerdán [2011], who studied the effect of knowledge management on innovation and corporate performance in 310 Spanish organizations. The results showed that knowledge management impacts innovation and organizational performance, both directly and indirectly through the improvement of innovation capability. Specifically, the study shows how codification and personalization strategies have a significant positive impact on financial results. In the same regard, Darroch [2005] studied the link between knowledge management, innovation, and performance. The results demonstrate that all three knowledge management components (knowledge acquisition, knowledge dissemination, and responsiveness to knowledge) have a direct effect on innovation. Nonetheless, only responsiveness to knowledge shows a positive relationship with financial performance.

Also, cultural and firm environmental factors are relevant to innovation and firm performance. For instance, Prajogo and Ahmed [2006] conducted a study of 194 Australian firms that demonstrated the relationships between innovation stimulus and innovation capacity and between innovation capacity and innovation performance are significant and strong. The authors stress the need to develop the behavioral and cultural context and practices for innovation (stimulus) so that the firm can develop innovative capacity in R&D and technology and thus be able to achieve strong innovation outcomes and performance. In the same vein, Hogan and Coote [2014] investigated 100 law firms and found that layers of organizational culture, particularly norms, artifacts, and innovative behaviors, partially mediate the effects of values that support innovation on measures of firm performance.

Rosenbusch *et al.* [2011] studied the innovation–performance relationship in small and medium-sized businesses. The findings confirm that innovation orientation and activities create value for new and established businesses. The authors point out that "although innovation can imply high initial and continuous investments, risks, and uncertainty, the benefits such as differentiation from competition, customer loyalty, price premiums for innovative products, and entry barriers for potential imitators generally seem to outweigh the costs" [Rosenbusch *et al.*, 2011, p. 452]. Similarly, focusing on small and medium-sized business, Varis and Littunen [2010] found that the introduction of new product, process, and market innovations is positively associated with the growth of a firm, whereas none of the innovation types studied revealed a positive relationship with the firm profitability.

REFERENCES

Artz, K. W., Norman, P. M., Hatfield, D. E., and Cardinal, L. B. [2010]. A longitudinal study of the impact of R&D, patents, and product innovation on firm performance, *Journal of Product Innovation Management*, 27(5), pp. 725–740.

Atalay, M., Anafarta, N., and Sarvan, F. [2013]. The relationship between innovation and firm performance: An empirical evidence from Turkish automotive supplier industry, *Procedia-Social and Behavioral Sciences*, 75, pp. 226–235.

Audretsch, D. B. [2014]. From the entrepreneurial university to the university for the entrepreneurial society, *The Journal of Technology Transfer*, 39(3), pp. 313–321.

Avlonitis, G. J., and Salavou, H. E. [2007]. Entrepreneurial orientation of SMEs, product innovativeness, and performance, *Journal of Business Research*, 60(5), pp. 566–575.

Basberg, B. L. [1987]. Patents and the measurement of technological change: A survey of the literature, *Research Policy*, 16(2–4), pp. 131–141.

Bettis, R. A., and Hitt, M. A. [1995]. The new competitive landscape, *Strategic Management Journal*, 16(S1), pp. 7–19.

Calabrese, A., Campisi, D., Capece, G., Costa, R., and Di Pillo, F. [2013]. Competiveness and innovation in high-tech companies: An application to the Italian biotech and aerospace industries, *International Journal of Engineering Business Management*, 5(40), pp. 1–11.

Calantone, R. J., Cavusgil, S. T., and Zhao, Y. [2002]. Learning orientation, firm innovation capability, and firm performance, *Industrial Marketing Management*, 31(6), 515–524.

Cho, H. J., and Pucik, V. [2005]. Relationship between innovativeness, quality, growth, profitability, and market value, *Strategic Management Journal*, 26(6), pp. 555–575.

Coad, A., and Rao, R. [2008]. Innovation and firm growth in high-tech sectors: A quantile regression approach, *Research Policy*, 37(4), pp. 633–648.

Damanpour, F., and Evan, W. M. [1984]. Organizational innovation and performance: The problem of "organizational lag", *Administrative Science Quarterly*, pp. 392–409.

Darroch, J. [2005]. Knowledge management, innovation and firm performance, *Journal of Knowledge Management*, 9(3), pp. 101–115.

DeCarolis, D. M., and Deeds, D. L. [1999]. The impact of stocks and flows of organizational knowledge on firm performance: An empirical investigation of the biotechnology industry, *Strategic Management Journal*, 20(10), pp. 953–968.

Demirel, P., and Mazzucato, M. [2012]. Innovation and firm growth: Is R&D worth it? *Industry and Innovation*, 19(1), pp. 45–62.

Dosi, G. [1988]. Sources, procedures, and microeconomic effects of innovation, *Journal of Economic Literature*, pp. 1120–1171.

Ernst, H. [2001]. Patent applications and subsequent changes of performance: Evidence from time-series cross-section analyses on the firm level, *Research Policy*, 30(1), pp. 143–157.

García-Manjón, J. V., and Romero-Merino, M. E. [2012]. Research, development, and firm growth. Empirical evidence from European top R&D spending firms, *Research Policy*, 41(6), pp. 1084–1092.

Garcia, R., and Calantone, R. [2002]. A critical look at technological innovation typology and innovativeness terminology: A literature review, *Journal of Product Innovation Management*, 19(2), pp. 110–132.

Geroski, P., Machin, S., and Van Reenen, J. [1993]. The profitability of innovating firms, *The RAND Journal of Economics*, pp. 198–211.

Gopalakrishnan, S., and Damanpour, F. [1997]. A review of innovation research in economics, sociology and technology management, *Omega*, 25(1), pp. 15–28.

Griliches, Z., Pakes, A., and Hall, B. H. [1986]. The value of patents as indicators of inventive activity, *NBER Working Paper*, No. 2083.

Gunday, G., Ulusoy, G., Kilic, K., and Alpkan, L. [2011]. Effects of innovation types on firm performance, *International Journal of Production Economics*, 133(2), pp. 662–676.

Hagedoorn, J., and Cloodt, M. [2003]. Measuring innovative performance: Is there an advantage in using multiple indicators? *Research Policy*, 32(8), pp. 1365–1379.

Han, J. K., Kim, N., and Srivastava, R. K. [1998]. Market orientation and organizational performance: Is innovation a missing link? *Journal of Marketing*, 62(4), pp. 30–45.

Hanel, P., and St. Pierre, M. [2006]. Industry–university collaboration by Canadian manufacturing firms, *The Journal of Technology Transfer*, 31(4), pp. 485–499.

Hogan, S. J., and Coote, L. V. [2014]. Organizational culture, innovation, and performance: A test of Schein's model, *Journal of Business Research*, 67(8), pp. 1609–1621.

Jiménez-Jiménez, D., and Sanz-Valle, R. [2011]. Innovation, organizational learning, and performance, *Journal of Business Research*, 64(4), pp. 408–417.

Kande, A. W., Kirira, P. G., and Michuki, G. N. [2017]. University Industry Collaboration and Innovativeness of Firms: Evidence from Kenya Innovation Survey, *International Journal of Innovation Education and Research*, 5(3), pp. 1–10.

Keskin, H. [2006]. Market orientation, learning orientation, and innovation capabilities in SMEs: An extended model, *European Journal of Innovation Management*, 9(4), pp. 396–417.

Lawless, M. W., and Anderson, P. C. [1996]. Generational technological change: Effects of innovation and local rivalry on performance, *Academy of Management Journal*, 39(5), pp. 1185–1217.

Lieberman, M. B., and Montgomery, D. B. [1988]. First-mover advantages, *Strategic Management Journal*, 9(S1), pp. 41–58.

López-Nicolás, C., and Meroño-Cerdán, Á. L. [2011]. Strategic knowledge management, innovation and performance, *International Journal of Information Management*, 31(6), pp. 502–509.

Lumpkin, G. T., and Dess, G. G. [1996]. Clarifying the entrepreneurial orientation construct and linking it to performance, *Academy of Management Review*, 21(1), pp. 135–172.

Mann, R. J., and Sager, T. W. [2007]. Patents, venture capital, and software startups, *Research Policy*, 36(2), pp. 193–208.

Mansury, M. A., and Love, J. H. [2008]. Innovation, productivity and growth in US business services: A firm-level analysis, *Technovation*, 28(1–2), pp. 52–62.

Murphy, G. B., Trailer, J. W., and Hill, R. C. [1996]. Measuring performance in entrepreneurship research, *Journal of Business Research*, 36(1), pp. 15–23.

Narin, F., Noma, E., and Perry, R. [1987]. Patents as indicators of corporate technological strength, *Research Policy*, 16(2–4), pp. 143–155.

OECD [2005]. *Oslo Manual 2005* (OECD Publishing, France).

OECD [2018]. *Oslo Manual 2018* (OECD Publishing, France).

Pelham, A. M. [2000]. Market orientation and other potential influences on performance in small and medium-sized manufacturing firms, *Journal of Small Business Management*, 38(1), pp. 48–67.

Prajogo, D. I., and Ahmed, P. K. [2006]. Relationships between innovation stimulus, innovation capacity, and innovation performance, R*&D Management*, 36(5), pp. 499–515.

Roberts, P. W. [1999]. Product innovation, product–market competition and persistent profitability in the US pharmaceutical industry, *Strategic Management Journal*, 20(7), pp. 655–670.

Roper, S., and Love, J. H. [2002]. Innovation and export performance: Evidence from the UK and German manufacturing plants, *Research Policy*, 31(7), pp. 1087–1102.

Rosenbusch, N., Brinckmann, J., and Bausch, A. [2011]. Is innovation always beneficial? A meta-analysis of the relationship between innovation and performance in SMEs, *Journal of business Venturing*, 26(4), pp. 441–457.

Rothwell, R. [1994]. Towards the fifth-generation innovation process, *International Marketing Review*, 11(1), pp. 7–31.

Salavou, H. [2002]. Profitability in market-oriented SMEs: Does product innovation matter? *European Journal of Innovation Management*, 5(3), pp. 164–171.

Shaukat, S., Nawaz, M. S., and Naz, S. [2013]. Effects of innovation types on firm performance: An empirical study on Pakistan's manufacturing sector, *Pakistan Journal of Commerce and Social Sciences* (*PJCSS*), 7(2), pp. 243–262.

Schumpeter, J. [1934]. *The Theory Of Economic Development*, 1st edn. (Harvard University Press, Cambridge).

Schumpeter, J. [1942]. *Capitalism, Socialism and Democracy*, 1st edn. (Harper, New York).

Sohn, D. W., Hur, W., and Kim, H. J. [2010]. Effects of R&D and patents on the financial performance of Korean venture firms, *Asian Journal of Technology Innovation*, 18(2), pp. 169–185.

Terziovski, M. [2010]. Innovation practice and its performance implications in small and medium enterprises (SMEs) in the manufacturing sector: A resource-based view, *Strategic Management Journal*, 31(8), 892–902.

Therrien, P., Doloreux, D., and Chamberlin, T. [2011]. Innovation novelty and (commercial) performance in the service sector: A Canadian firm-level analysis, *Technovation*, 31(12), pp. 655–665.

Thornhill, S. [2006]. Knowledge, innovation and firm performance in high-and low-technology regimes, *Journal of Business Venturing*, 21(5), pp. 687–703.

Varis, M., and Littunen, H. [2010]. Types of innovation, sources of information and performance in entrepreneurial SMEs, *European Journal of Innovation Management*, 13(2), pp. 128–154.

Venkatraman, N., and Ramanujam, V. [1986]. Measurement of business performance in strategy research: A comparison of approaches, *Academy of Management Review*, 11(4), pp. 801–814.

Weerawardena, J. [2003]. Exploring the role of market learning capability in competitive strategy, *European Journal of Marketing*, 37(3–4), pp. 407–429.

Wolff, J. A., and Pett, T. L. [2006]. Small-firm performance: Modeling the role of product and process improvements, *Journal of Small Business Managem*ent, 44(2), pp. 268–284.

INDEX

U

V

CPSIA information can be obtained
at www.ICGtesting.com
Printed in the USA
BVHW042354150320
574806BV00006B/14